Intelligent Exploration in Sustainable Design

Jiawei Yao

Intelligent Exploration in Sustainable Design

2021–2024 DigitalFUTURES Workshop Achievements

Springer

Jiawei Yao
College of Architecture and Urban Planning
Tongji University
Shanghai, China

ISBN 978-981-96-6742-0 ISBN 978-981-96-6743-7 (eBook)
https://doi.org/10.1007/978-981-96-6743-7

This work was supported by National Natural Science Foundation of China (52278041 and 52338004).

This Springer imprint is published by the registered company Springer Nature Singapore Pte Ltd.
The registered company address is: 152 Beach Road, #21-01/04 Gateway East, Singapore 189721, Singapore

Competing Interests The author has no competing interests to declare that are relevant to the content of this manuscript.

Introduction: Disciplinary Awakening in an Era of Crisis

In the twenty-first century, human society is facing a dual crisis: ecological collapse and technological alienation. Climate change, resource depletion, and the exponential evolution of digital technology have pushed architecture to an unprecedented crossroads—should it remain confined to the traditional domain of "form and function," or embrace disruptive, interdisciplinary transformations? Against this backdrop, the DigitalFUTURES workshop platform—one of the most influential initiatives in the global digital architecture field—has led a series of workshops from 2021 to 2024. At this event, SDII team completed four workshops. These workshops, structured around four key themes—"Environmental Intelligence, Carbon-Neutral Cities, Morphcarbon Causality, and Intelligent Reconstruction of Multiphysical Fields"—constitute a systematic experimental framework responding to the global climate crisis. These themes not only signify innovations in technological tools but also pose fundamental philosophical inquiries into the essence of architecture: How should architecture redefine its knowledge system, methodology, and social mission amid the twin forces of digitalization and sustainability?

This chapter will delve into the internal logic, technological pathways, and disciplinary significance of these 4 years of exploration, revealing how the workshop-based pedagogical model has propelled architecture from the "art of making" to the "science of governance."

Thematic Evolution: From Tool Innovation to Systemic Revolution

1. 2021: Environmental Artificial Intelligence in Design
   - Data-Driven Ecological Awakening
2. 2022: AI For Carbon Neutral City
   - Data-Driven Ecological Awakening

3. 2023: Decoding Morphcarbon Causality
   - The Ecological Judgment of Aesthetics
4. 2024: Intelligent Reconstruction of Multiphysical Fields
   - Taming Complexity

Logical Progression: The Fourfold Transition from Tools to Paradigm

1. The Spiral Ascent of Technological Depth
   - Data Complexity: From singular environmental data (2021) to the integration of multiphysical field data (2024).
   - Algorithmic Capability: From passive response (environmental feedback) to active generation (AI-driven creation).
   - System Scale: From individual buildings (2021) to urban systems (2022) and ultimately to cross-scale coupling (2024).
2. The Disruptive Iteration of Disciplinary Paradigms
   - Epistemology: From "form follows function" to "form follows carbon flows" (2023), and eventually to "form tames complexity" (2024).
   - Methodology: A shift from empirical induction to a "hypothesis-model-validation" scientific research paradigm.
   - Values: From the pursuit of aesthetic uniqueness to a triadic responsibility framework integrating ecology, humanities, and technology.
3. The Strategic Upgrade of Architecture's Societal Role
   - Architects as System Coordinators: Managing multiple systems, including environmental data, carbon metabolism, and physical performance, thereby transitioning from "spatial creators" to "complexity managers."
   - Architecture as a Socio-Technological Interface: Serving as an intermediary platform connecting policy (e.g., carbon neutrality goals), technology (e.g., AI), and the public (e.g., participatory design).

Conclusion: Reconstructing Architecture's "Noah's Ark"

The 4-year pedagogical outcomes of the SD-II workshops reveal a profound truth: In the face of climate crises and the digital wave, architecture must become a modern-day "Noah's Ark"—one that harnesses cutting-edge technological engines while safeguarding the ecological and humanistic foundations of civilization. The blueprint for this ark lies within the codes, models, and narratives of these workshops. The mission of this book is to translate these blueprints into actionable strategies, empowering every reader to become a "future architect" on this ark—one who wields algorithms in one hand and humanity in the other, navigating toward an uncertain yet hopeful new frontier where digitalization and sustainability harmonize.

Dr. Jiawei Yao

Contents

Part III DigitalFUTURES 2023: Decoding Morphcarbon Causality

Tutor Information

Jiawei Yao is an associate professor in the Department of Architecture at the College of Architecture and Urban Planning, Tongji University. He is funded by Shanghai Sailing Talent Program and serves as the Vice Director of Shanghai Engineering Research Center of Digital Building Construction and the Deputy Secretary-General of the Digital Architecture Branch of Shanghai Architecture Society. He is also a member of the Academic Committee of Computational Design of the Architectural Society of China and DigitalFUTURES. Yao has led several national and provincial-level research projects, including the National Natural Science Foundation of China and the Postdoctoral Fund of China. He is a guest editor for several international and domestic journals, such as Building and Environment, and has participated in the drafting and release of several local standards in Shanghai.

Minggang Yin currently at the Shanghai Architectural Science and Technology Innovation Center of East China Architectural Design and Research Institute Co., Ltd., is a senior architectural performance analyst. With extensive experience in green building energy-saving design consultation and specialized architectural performance analysis, he has contributed to numerous landmark projects. Yin is well-versed in both domestic and international architectural performance-based design methods and has published papers in core journals within the field. He actively advocates for the integration of advanced methods such as BIM technology, parametric analysis, machine learning, and VR/AR into green and low-carbon building design.

Chenyu Huang is the principal researcher at SD-II Lab. He is a Ph.D. student at the College of Architecture and Urban Planning, Tongji University, and a recipient of the CAADRIA Young Award. His research primarily focuses on AI for Science, optimization of urban built environments, and urban building energy modeling. Huang has published over 20 papers in renowned domestic and international journals, including Sustainable Cities and Society, Building and Environment, Heliyon (journal affiliated with Cell), and Architectural Journal. He has received the Outstanding Paper Award at the Tsinghua University Ph.D. Forum and the Best Paper Award at the 5th International Conference on Building Energy and Environment.

Dr. Jing Meng a tenured Associate Professor at the Bartlett School of Sustainable Architecture, University College London, and a member of the Cambridge Centre for Environment, Energy, and Natural Resource Governance (C-EENRG), serves as an Associate Editor for the SCI journal, Journal of Cleaner Production. Her research focuses on the theory and modeling of climate change and energy policy. Dr. Meng has published multiple papers in internationally renowned journals such as Nature Communications, Nature Climate Change, PNAS, and One Earth. She has been recognized as a "Highly Cited Researcher" by Clarivate Analytics in 2020 and 2021, and was honored with the Emerging Sustainability Leader Award in 2020, as well as the Nature Communications Top 50 Earth and Planetary Science Papers Award.

Bin Fu the Regional Director for China at METEODYN, a subsidiary of CLS Group, has over 15 years of experience in research and practice in the field of renewable energy. His expertise includes the design and implementation of intelligent digital platforms for renewable energy, fluid and meteorological modeling, data analysis, and power forecasting. He has been responsible for completing system design and implementation for energy groups such as CGN Wind Power, Three Gorges New Energy, China Resources Power, and State Energy Investment Group, as well as large-scale energy equipment manufacturers. CLS Group, founded by the French Space Agency in 1986, provides cutting-edge technologies in Earth observation, meteorological and oceanic simulation, data fusion, and artificial intelligence.

Yupeng Wu is a Professor of Building Physics at the Department of Architecture and Built Environment, the University of Nottingham, UK. His researches focus on building energy simulation and integration of renewable energy systems into buildings and built environment. He has been awarded ~ £4m in research grants including funding from the Engineering and Physical Sciences Research Council UK (EPSRC), Innovate UK, etc. He has published over 100 peer-reviewed papers in SCI journals and conference proceedings contributing to the state of the art in renewable energy technologies and their application to the built environment. Jack serves as subject Editor for the journal of 'Renewable Energy', asscoiated Editor for Arcthitectural Intellegence and Editorial board member for Building Simulation. He is a full member of the EPSRC peer reviewer college and is also reviewer for European, Canadian and US funding agencies.

Sergio Altomonte Professor of Building Physics at KU Leuven, leading the Building and Climate Research Group. A member of the Passive Low Energy Architecture (PLEA), an international organization dedicated to global discussions on sustainable building and urban design. Fellow of the Higher Education Academy (FHEA).

Maider Llaguno-Munitxa Assistant Professor at the Laboratory for Architecture and the Built Environment (LAB) at KU Leuven and Head of the Urban Science, Climate, and Health team within the Architecture and Climate Group at UCLouvain. Her research focuses on data science and digital design tools to develop innovative design practices for improving urban environmental quality and health.

Part I
DigitalFUTURES 2021: Environmental Artificial Intelligence in Design

Mentors
Yao Jiawei, Ying Minggang, Huang Chenyu, Zhang Gengjia

Workshop Participants
Gao Xiangyu, Yan Hainan, Xiao Tong, Zhou Bingrui, Yang Ye'an, Fang Yikun, Li Jingyi, Li Shuaijie, Zheng Zhongyi, Wang Xiaoxin, Lian Zhiyuan, Guo Mingyue, Jiang Qianli, Zhang Yulin, Yan Lina, Dong Yuxiang, Ji Hongfei, Sui Yingda, Shao Yiming, Zhou Zhiwei, Liu Deli, Chen Cuilan, Xu Shiyi, Xu Ke, Li Bin, Xia Tianxiang, He Pan, Lin Duobin, You Baihui, Jia Jing, Hu Beiyu, Cui Yuechen, Lu Yijun, Feng Wenhan, Tu Han, Buyin, Wang Yayu, He Simian, Li Zehao, He Weili, Wang Chidi, Tang Xu, Ji Yunzhu, Hao Tongping, Peng Huiwang, Wang Hongxin, Zheng Yun, Gan Xinran, Duan Shufan, Zha Xinyu, Huang Tingting.

Topic
The emergence of Environmental Performance-Based Design aims to provide solutions and strategies for cities and buildings to adapt to climate change and offer more comfortable living spaces. Traditionally, environmental performance has been considered only as a conceptual design guideline and evaluation factor. Adjustments to spatial forms have mainly relied on designers and consultants collaborating in a "black-box" and experiential passive design approach. In contrast, Environmental Performance-Driven Design, spurred by rapidly advancing artificial intelligence tools, is a proactive design approach. It emphasizes selecting environmental performance as a goal in the early stages of design, guiding the generation of design solutions through simulation analysis, feedback of results, and iterative optimization.

Content
The workshop will focus on key theories and technical issues surrounding Environmental Performance-Driven Generative Design. Together with participants, we will explore how to integrate environmental performance evaluation and generative design, and investigate the broader applications of artificial intelligence in optimizing environmental performance. The workshop will combine theoretical teaching with technical guidance to help participants understand the critical issues and solutions

within the framework of Environmental Performance-Driven Generative Design. Participants will engage in thematic research in small groups based on the course content, and the outcomes of the workshop will include posters, 2–3 min videos, and presentations.

Trainee Information in 2021

Total applicants: 150

Number of officially admitted trainees: 52

Admission rate: 35%

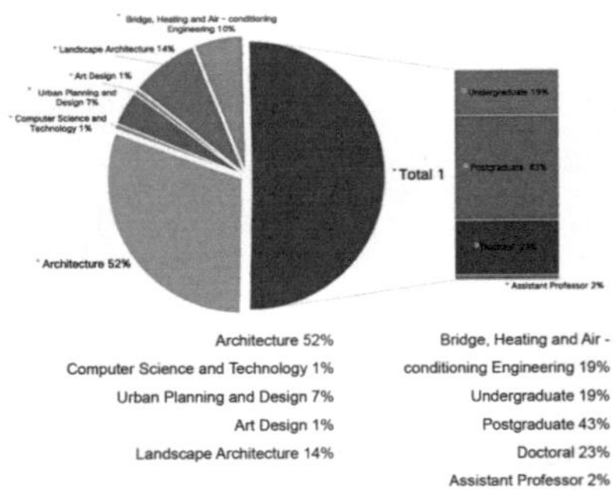

Paper publication

·Yao Jiawei, Huang Chenyu, Yuan Feng. Research on the Intelligent Generative Design Method of Architecture Driven by Multi - environmental Substances [J]. Time + Architecture, 2021, (06): 38 - 43. DOI: 10.13717/j.cnki.ta.2021.06.009.

·Ji Yunzhu, Yao Jiawei, Hao Tongping, et al. Research on the Intelligent Generative Method of Residential Areas Oriented by Machine Learning and Environmental Performance Optimization [J]. Architecture Technique, 2023, 29(07): 70 - 74. DOI: 10.19953/j.at.2023.07.022.

·Chen Jinmin, Yan Hainan. Research on the Building Layout at the Block Scale Based on Wind Environment Optimization - Taking Guangzhou, a City in the Hot - humid Region, as an Example [J]. Architecture Technique, 2021, 27(09): 73 - 77. DOI: 10.19953/j.at.2021.09.008.

·Huang Chenyu, Zhang Gengjia, Zhang Xingzhao, et al. Energy - Driven Urban Intelligent Generation - Regional Energy Consumption Optimization Design in the Early Design Stage [J]. Architecture Technique, 2021, 27(09): 78 - 83. DOI: 10.19953/j.at.2021.09.009.

·Wang Xiaoxin, Li Jingyi. Research on the Daylighting Optimization of the Residential Area in Hankou Riverside Based on the Generative Adversarial Network (GAN) and the NSGA - 2 Genetic Algorithm [J]. Architecture Technique, 2021, 27(09): 84 - 88. DOI: 10.19953/j.at.2021.09.010.

Chapter 1
Shared Office 2.0 of Post-epidemic Era

Generate the Optimal Solution of Plane with the Goal of Lower Lighting Energy Consumption in the Initial Design Stage

Members
Gao Xiangyu, Yan Hainan, Xiao Tong, Zhou Bingrui, Yang Ye'an, Fang Yikun

Background

Tracing office architecture evolution through capitalist development reveals increasing boredom and waste, prioritizing economy and efficiency over human perspective. Contemporary focus on shared offices and dynamic floor plans heightens energy consumption. This study seeks an optimal layout balancing light and energy efficiency via a parametric approach.

Method

1. Building Plan Generation

In the post-epidemic era, indoor lighting is crucial for creating a healthy and comfortable shared office environment. The study analyzes sunlight hours and solar radiation intensity in Shanghai, New York, and Singapore. Based on an 8x8 grid, prototype office buildings with dimensions of 48 m by 32 m are constructed, and 15 feasible floor plans are selected by incorporating side courts or atriums.

2. Environmental Indicator Simulation

Ladybug and Honeybee plug-ins are utilized to analyze indoor lighting, measure sunlight satisfaction rates, and select the best atrium design. GHCpython and pedsim plug-ins simulate pedestrian flow, establish activity timetables, and determine behavior probabilities based on functional layouts (4:1:1 for meeting, communication, and leisure) and office events. Thermal simulations of human flow are conducted to estimate indoor energy consumption. Visual analysis of personnel movement informs optimization of building facades.

J. Yao, *Intelligent Exploration in Sustainable Design*,
https://doi.org/10.1007/978-981-96-6743-7_1

3. CNN Accelerated Calculation

A CNN model is developed to accelerate the design process using optimal parameters. The model inputs building plans and room divisions and outputs sunlight satisfaction rates. After comparing various machine learning models, a CNN regression model is chosen for training. Wallacei X multi-objective genetic algorithm optimizes both daylight satisfaction rates and lighting energy consumption, yielding an optimal solution for functional layout generation.

Significance
This study combines the current AI technology and tries to build a more humanized office space intelligent generation tool, through which only simple parameter inputs are needed to provide architects with more rational suggestions for pre-design planning.

1.1 Technology Method

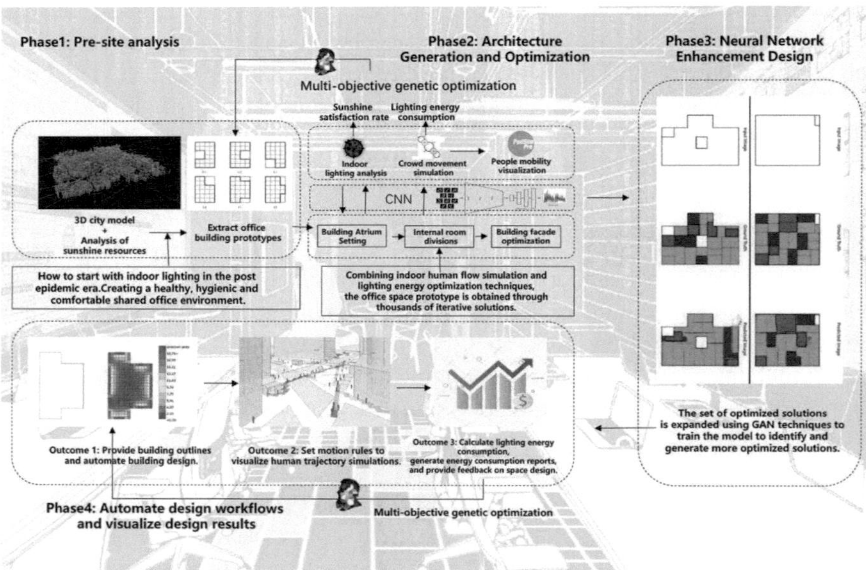

This diagram illustrates an AI-driven architectural design optimization process, encompassing four key phases: pre-site analysis, architecture generation and optimization, neural network enhancement design, and automated design visualization. In Phase 1 (Pre-site Analysis), 3D city modeling and sunlight resource analysis are utilized to extract office building prototypes, establishing design objectives for healthy post-pandemic office spaces. Phase 2 (Architecture Generation and Optimization) employs multi-objective genetic optimization and Convolutional Neural Networks (CNNs) to analyze indoor lighting, simulate crowd movement, and optimize architectural spaces. Phase 3 (Neural Network Enhancement Design) applies Generative Adversarial Networks (GANs) to expand the set of optimized solutions, enabling the AI model to identify and generate more refined design options. Finally, in Phase 4 (Automated Design and Visualization), crowd trajectory simulations and lighting energy consumption analysis are conducted to further optimize and validate the design. This research integrates deep learning, data-driven optimization, and multi-objective generative design, providing new possibilities for intelligent and sustainable office space design.

1.2 Sunlight Intensity Analysis

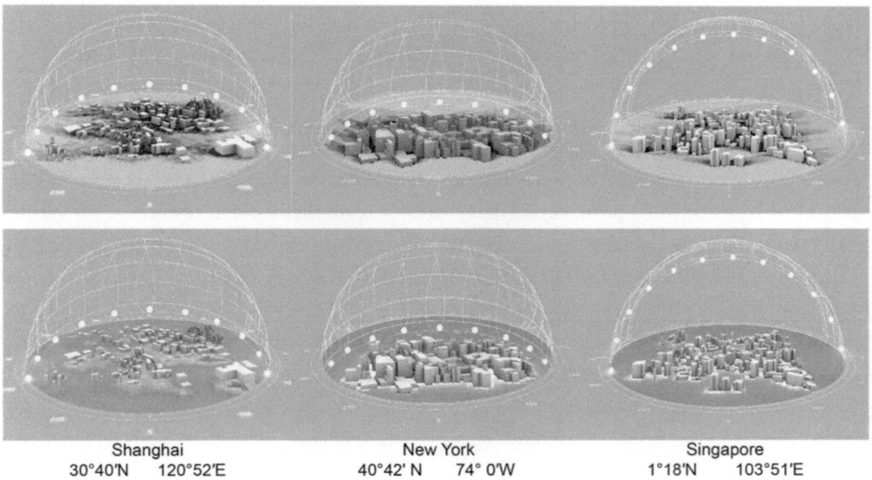

Solar radiation patterns are crucial for optimizing office spaces. The solar path and sunlight intensity in different cities determine key design factors such as daylighting strategies, lighting energy use, HVAC loads, and shading optimization. Integrating solar radiation analysis into AI-driven design enhances the sustainability of intelligent office spaces, creating efficient, healthy, and eco-friendly work environments for cities in diverse climates.

1.3 15 Types of Office Planes

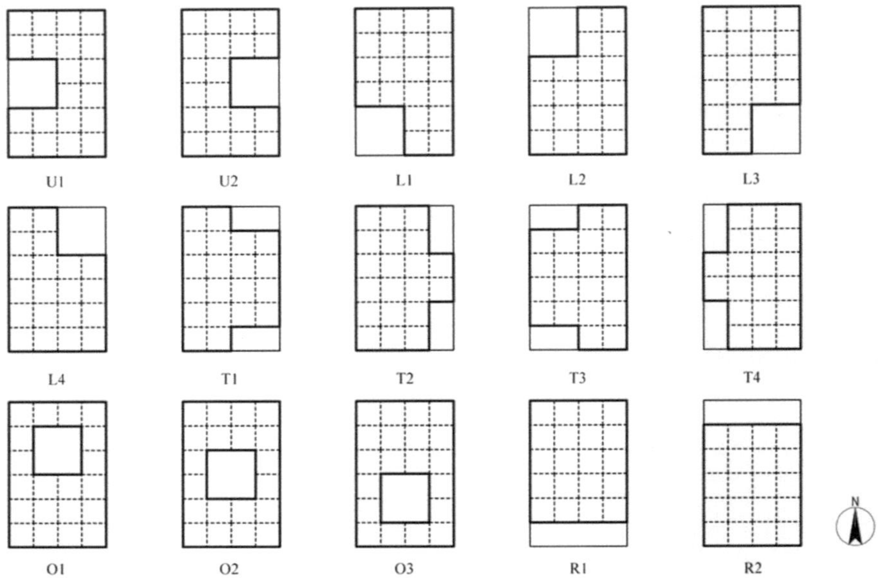

These office layout prototypes incorporate atriums or side courtyards to enhance natural daylighting, improve ventilation, and elevate workplace comfort. By strategically adjusting the placement and scale of openings in different building forms, each prototype optimizes indoor environmental quality according to specific spatial configurations. Atriums or side courtyards are introduced to improve daylight penetration, enhance natural ventilation, and create a more comfortable workspace. These prototypes serve as foundational models, providing diverse design possibilities that can be further optimized using AI-driven data analysis and performance-based simulations.

1.4 Interior Lighting and Atrium Setting

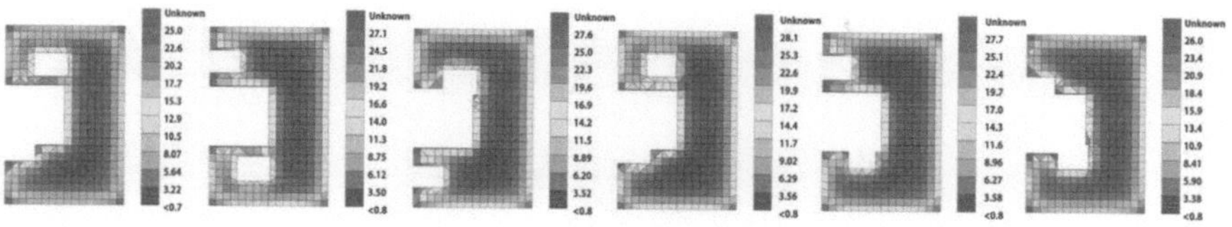

This image presents the results of daylight simulation for office space prototypes using the Ladybug plugin, generating datasets through batch simulation for machine learning or architectural optimization. Ladybug is an environmental analysis tool that calculates indoor daylight distribution based on meteorological data, solar radiation, and building geometry.

1.5 Internal Room Division Results

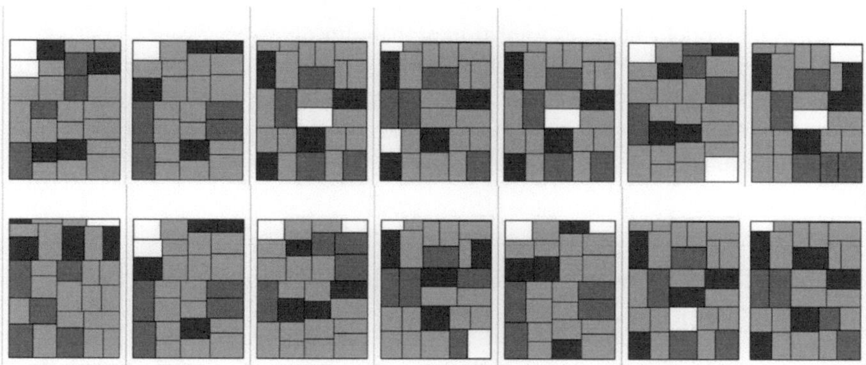

This image presents a dataset of internal room divisions in office spaces generated through automated algorithms or AI optimization. In these layouts, green represents office spaces, red indicates meeting rooms, blue denotes auxiliary spaces (such as lounges and pantries), and white corresponds to atriums or side courtyards.

1.6 Trajectory of Movement of People in the Building

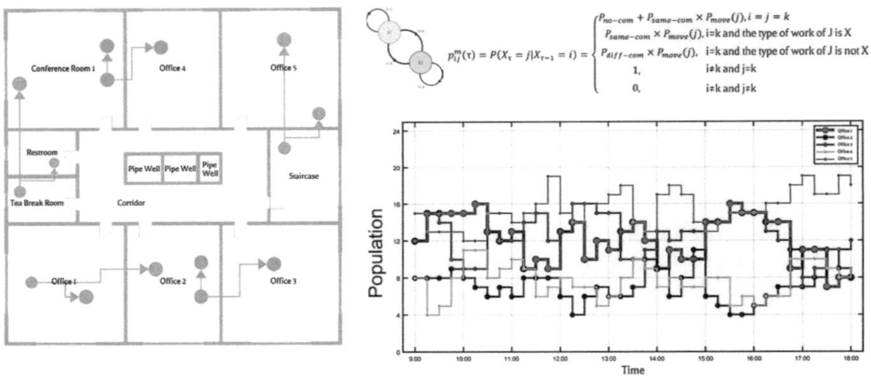

This diagram illustrates Markov Chain-based analysis of office space movement, studying how people move through the building over time. The left plan visualizes movement paths, with green and yellow dots marking starting points and arrows indicating direction, covering key areas like offices, meeting rooms, and break rooms. The top-right model calculates transition probabilities between spaces, considering intra-department, inter-department, and job-type influences to predict dynamic distribution. The bottom-right line graph shows occupancy changes in offices throughout the day (09:00–18:00), with the X-axis for time and Y-axis for headcount, reflecting density variations.

1.7 Impact of Personnel Movement on Lighting Energy Consumption

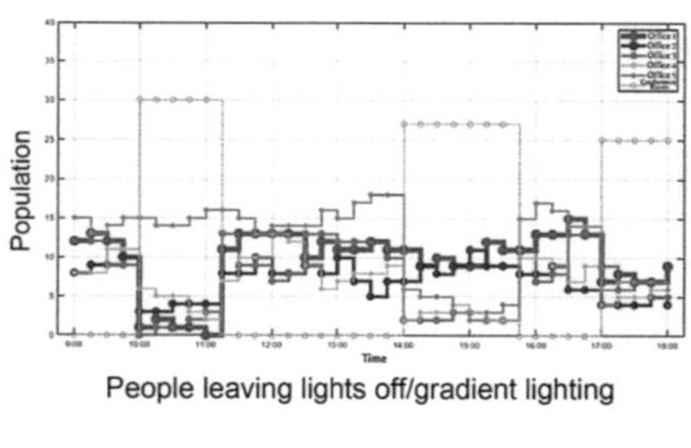

People leaving lights off/gradient lighting

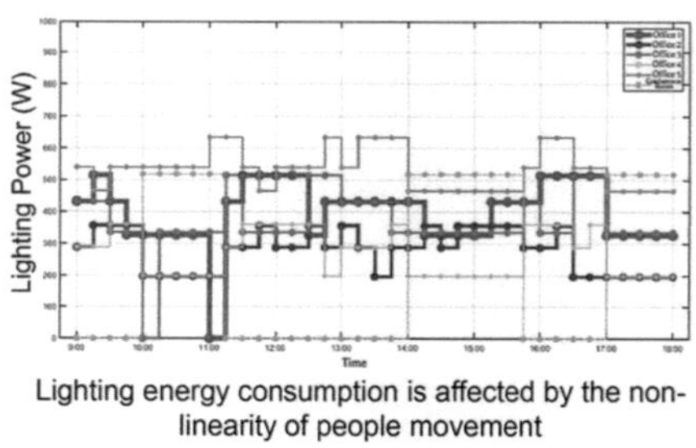

Lighting energy consumption is affected by the non-linearity of people movement

This image illustrates the impact of dynamic personnel movement on lighting energy consumption, revealing how the nonlinear patterns of occupant flow in office spaces influence lighting power variation. The line graph on the left shows fluctuations in the number of people in different office areas over time, while the graph on the right presents corresponding changes in lighting energy consumption. The study indicates that lighting consumption does not strictly follow a linear relationship with occupancy but is influenced by factors such as brief absences, lighting habits, and intelligent control systems.

1.8 Simulation of Human Flow

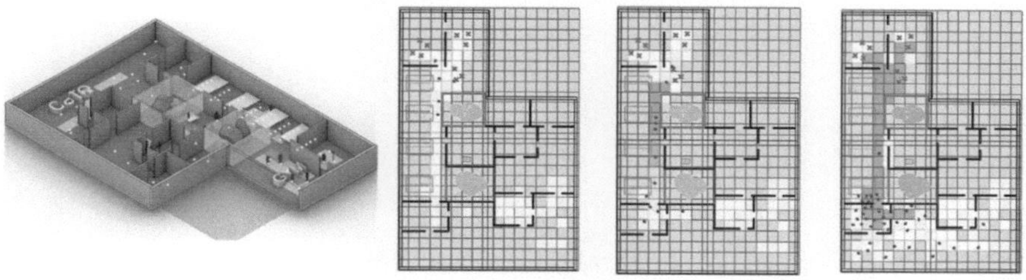

This diagram visualizes people flow simulation, analyzing movement patterns and space usage within a building. The left 3D model shows the office layout, including functional areas like meeting rooms, workstations, and common spaces. The right heatmap uses grids and color gradients to indicate crowd density, with red for high-traffic paths and blue for low-traffic zones. Over time, it reveals activity hotspots and flow trends, aiding in optimizing office circulation, space efficiency, and safety evacuation strategies.

1.9 Crowd View Simulation

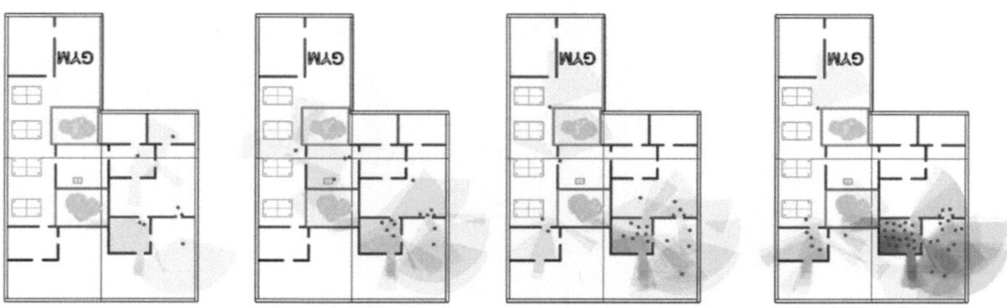

This image presents a Crowd View Simulation, designed to analyze the visibility range and spatial perception of individuals within a building. Each red dot represents an individual, while the semi-transparent gray fan-shaped areas indicate their field of view, showing the space they can observe.

1.10 Neural Network Accelerated Computing

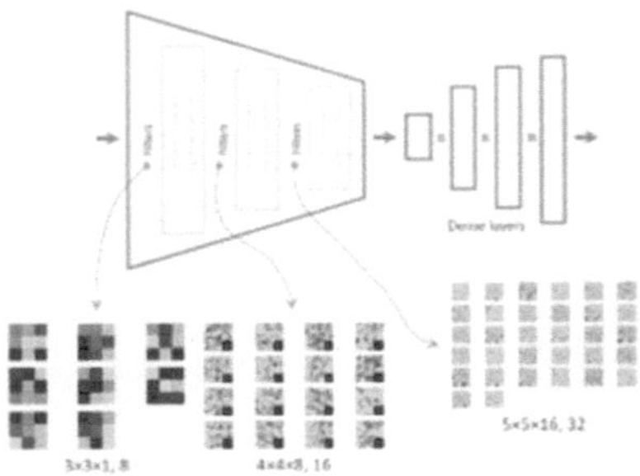

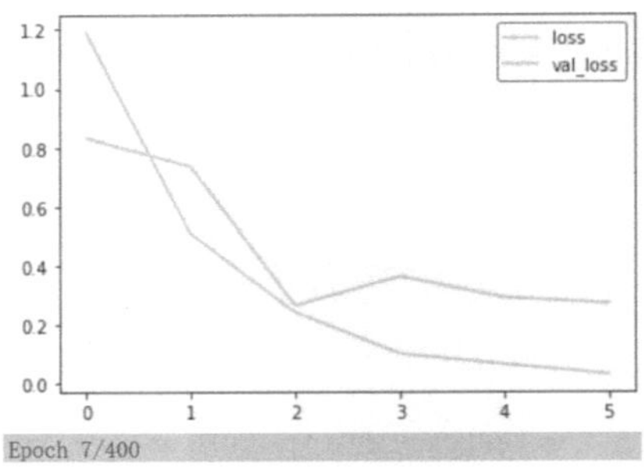

In this process, Convolutional Neural Networks (CNNs) accelerate office space optimization by learning the relationship between spatial division data and lighting and energy consumption metrics, enabling rapid building performance prediction. The CNN's input is a dataset of spatial divisions, including room layouts and functional zoning, while its output provides lighting and energy consumption indicators, replacing traditional simulations. Once trained, the CNN efficiently identifies high-quality layouts with optimal daylighting and low energy consumption, reducing computational costs and speeding up optimization. This approach makes AI-driven design smarter and more efficient, enabling exploration of optimal spatial configurations strategies.

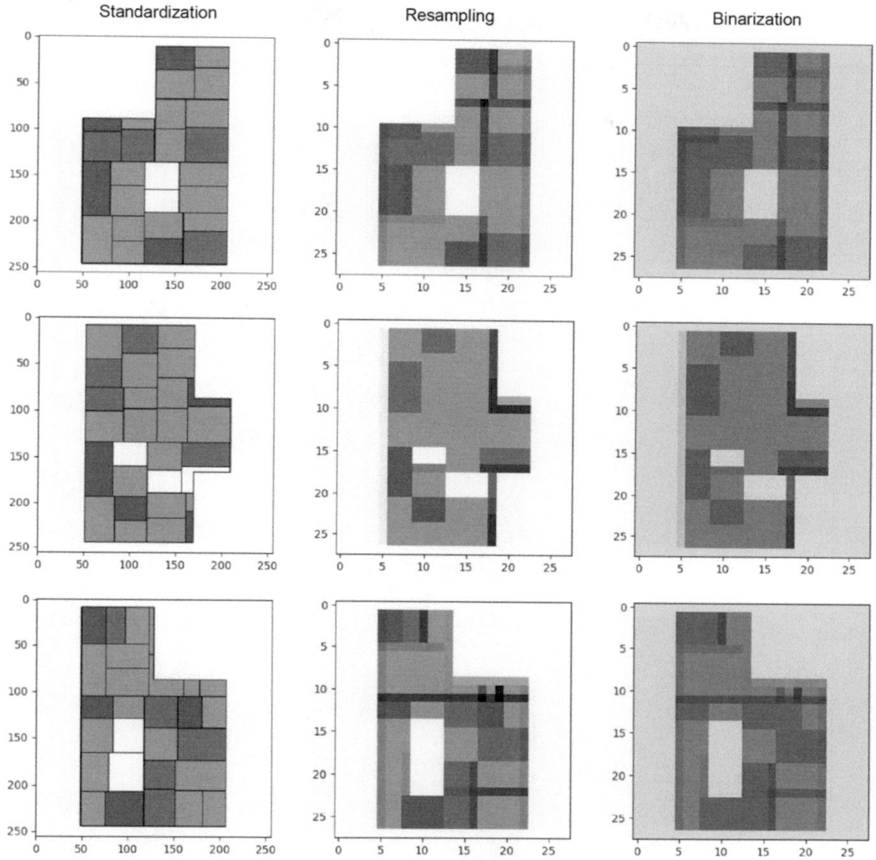

This image presents the standardization, resampling, and binarization of office space layout data, providing optimized data input for CNN model training.

1.11 Adversarial Generative Network Expansion

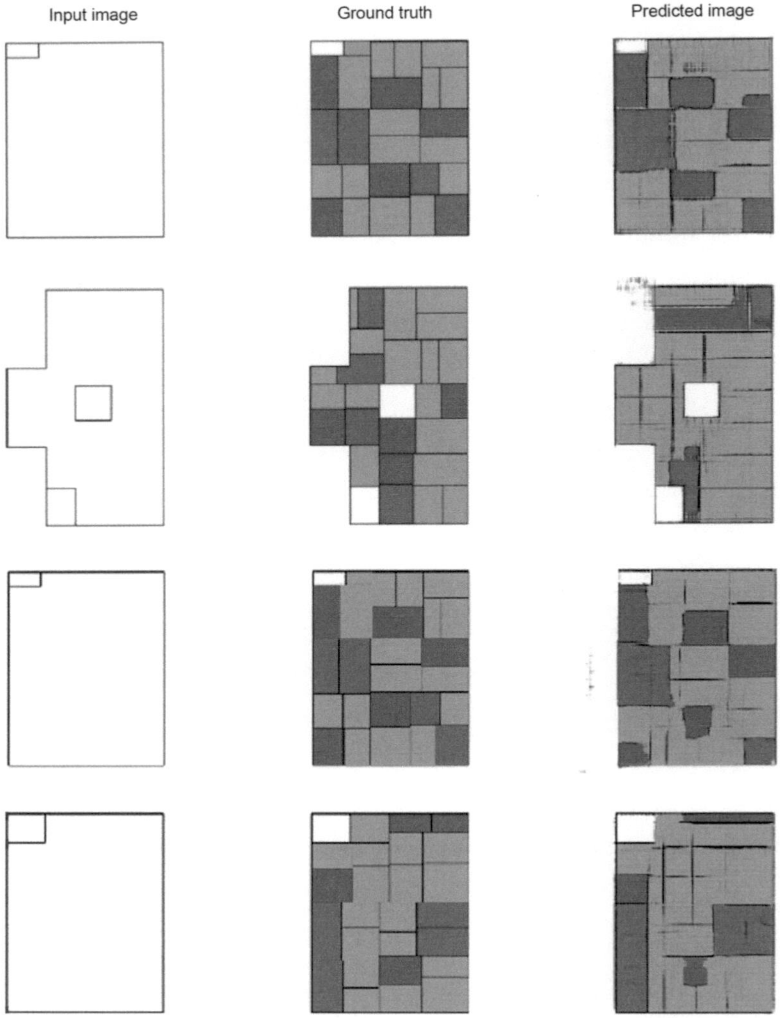

This section demonstrates the use of Pix2Pix Generative Adversarial Network (GAN) for automatic office space partitioning, generating high-performance internal room layouts based on input office floor boundaries. Pix2Pix is a deep learning model for image-to-image translation, capable of learning the mapping between input boundaries and spatial divisions. This method is trained on a high-quality floor plan dataset, enabling AI to efficiently predict spatial layouts and enhance office space optimization.

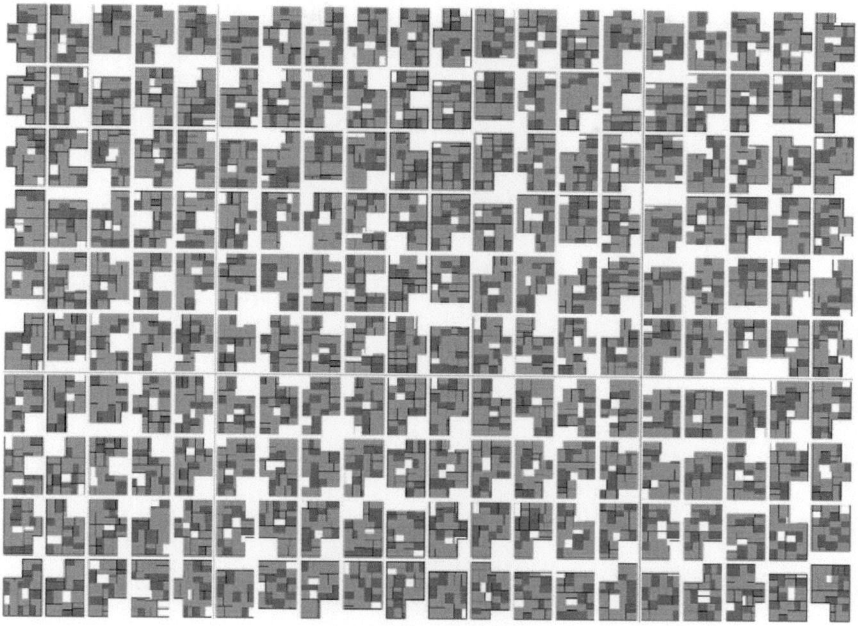

Predicted cases

This image showcases multiple high-quality office space layouts generated by Pix2Pix, where AI automatically produces rational internal room divisions based on the previously trained dataset. By learning spatial division patterns, AI enables large-scale and efficient automated design. These generated layouts can be utilized to optimize office space configurations, improve space utilization, and further incorporate daylighting and energy consumption metrics for selection, ultimately creating the most optimal office environment layouts.

Chapter 2
Simulation of Sunlight Environment and Optimization of Design Parameters in Urban Neighborhoods Based on Composite Adversarial Neural Network Generation Technology

Members
Li Jingyi, Li Shuaijie, Zheng Zhongyi, Wang Xiaoxin, Lian Zhiyuan

Background
This study investigates how urban block planning affects sunlight exposure. It aims to improve lighting conditions in high-latitude urban areas using scientific data. Current methods for designing roads and buildings in blocks rely on manual rules, which are laborious and struggle with complexity. The focus is on generating 3D urban models using machine learning, particularly Pix2Pix, to extract and learn from existing urban textures.

Method
1. Rules for Generating Blocks

The project focuses on Amsterdam, where sunlight exposure varies greatly between winter and summer. Sunlight significantly affects human health. Amsterdam's urban texture and waterway network are closely related, making it an ideal study site. Using parametric software, we process urban texture slices to extract block grid volume ratios, visualized as heatmaps. Road networks and nodes are categorized based on intersections and relationships (green, yellow, red). GAN is trained over 150 iterations to predict block volume ratios and road network distributions, with a dataset of 700 samples.

2. Model Training for Prediction

The parameterization generates distribution maps of the highest and lowest volume ratios and road node maps. The model predicts based on these two maps, allowing the machine to learn how to generate a complete block distribution map and a complete

J. Yao, *Intelligent Exploration in Sustainable Design*,
https://doi.org/10.1007/978-981-96-6743-7_2

road network using these predefined points and the road nodes. Next, Grasshopper is used to recognize and interpret the predicted maps, forming a 3D model of the city. Finally, sunlight simulation optimization is conducted, and the results of each generation of genetic algorithm are compared for subsequent selection.

3. Iteration

Sunlight simulation optimizes the winter solstice sunlight index, block network area ratio, and block building density.

Wallacei X optimization drives the generation of new volume ratios and road node maps for each iteration.

Significance
The current focus of the machine learning-based urban block generation tool is optimizing sunlight exposure. In the future, it could also consider optimizing urban wind patterns and thermal comfort to improve its effectiveness. This tool can evaluate urban block planning in high-latitude areas and guide future planning efforts.

2.1 Technology Method

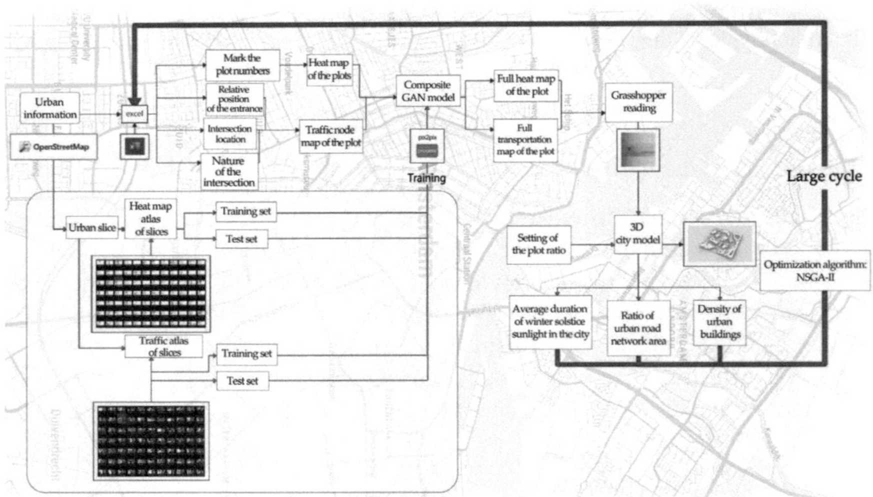

This technical framework integrates Generative Adversarial Networks (GANs) and the Non-dominated Sorting Genetic Algorithm II (NSGA-II) to develop a data-driven intelligent urban planning approach. First, urban data is extracted from OpenStreetMap, and heatmaps and traffic flow analysis are conducted for plots to form a training dataset. Then, Pix2Pix GAN is employed to predict urban spatial features, generating comprehensive plot heatmaps and transportation networks. These data are subsequently fed into Grasshopper for 3D city modeling, where key urban performance metrics such as solar exposure duration, road area ratio, and building density are computed. Finally, NSGA-II performs multi-objective optimization, automatically generating urban layouts that balance environmental, traffic, and building density factors. Through iterative cycles, this method continuously refines urban planning solutions, enhancing both efficiency and sustainability.

2.2 Study Area

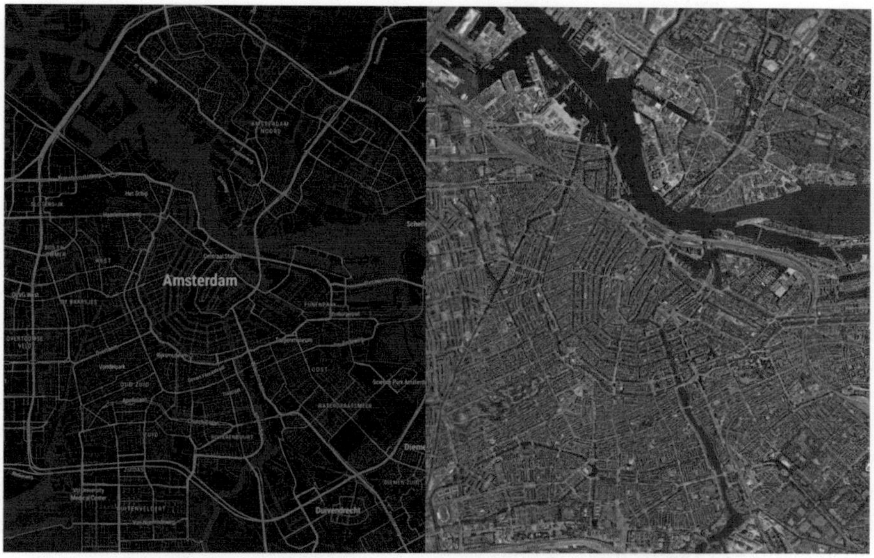

Amsterdam road network   Amsterdam satellite picture

This study selects Amsterdam as the research site, with the left image displaying its road network structure and the right image showing the satellite view, aimed at analyzing urban spatial layout, transportation networks, and optimization potential.

2.3 City Sample Generation Method

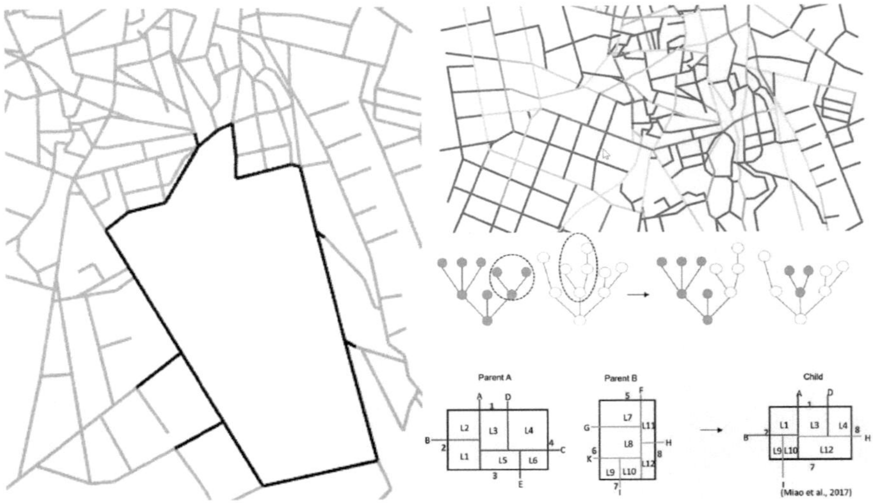

Previous city sample generation methods mainly relied on parametric modeling, data tree structures, or genetic algorithms. Parametric modeling creates urban layouts by adjusting parameters like road density and plot size, but its rule-based approach often lacks adaptability to complex urban forms. Data tree methods simulate spatial hierarchies and generate plots and road networks through recursive partitioning. While effective in organizing urban information, they depend heavily on predefined rules, limiting result diversity. Genetic algorithms use crossover and recombination to create new layouts but are constrained by fixed operational patterns, hindering the learning of real urban spatial patterns. This study adopts a data-driven generation approach, using deep learning and optimization strategies to directly learn spatial patterns from real urban data, improving flexibility and adaptability for more precise urban planning optimization.

2.4 Machine Learning Pix2Pix

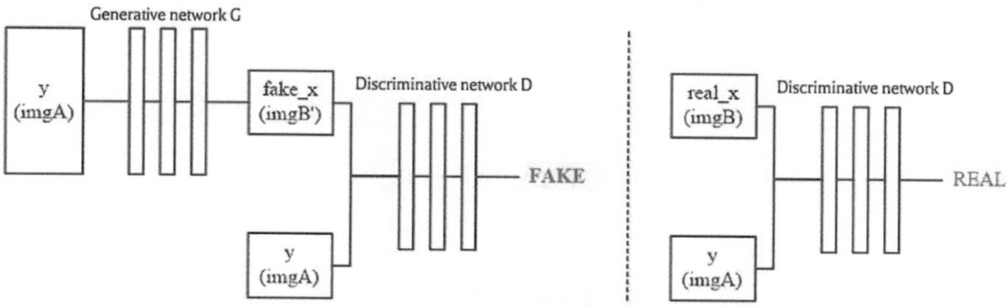

Pix2pix: The input end of the generation network G has only one input: condition y; it can be understood that condition y is an image im

Output: the generated result is regarded as imgB'

After the final training is completed: it is possible to transform one image A into another image B

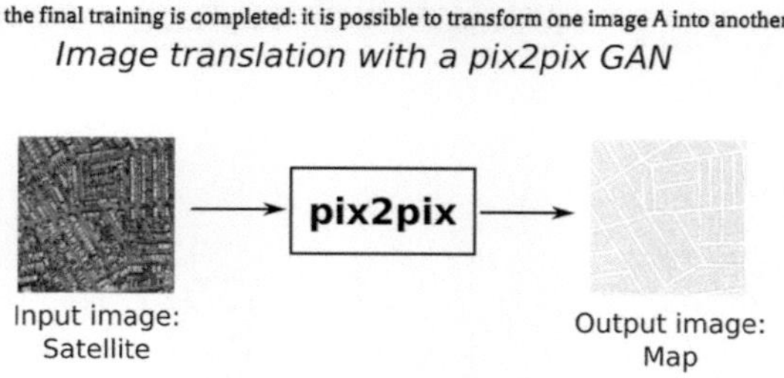

Pix2Pix is an image-to-image translation method based on Conditional Generative Adversarial Networks (cGANs). It consists of a generator (G) and a discriminator (D), where the generator learns to produce target images from input images, while the discriminator distinguishes between generated and real images, continuously optimizing the results to make them more realistic.

This study employs Pix2Pix to extract street networks, plot divisions, and other spatial structures from satellite imagery or urban data, enabling automatic urban spatial transformation. By learning from large-scale datasets, the model generates high-quality urban layouts, overcoming the limitations of traditional rule-based modeling.

Compared to parametric modeling or data tree methods, Pix2Pix eliminates the need for manually defined rules, instead learning transformation patterns directly from data. This approach enhances adaptability to diverse urban environments, automatically captures complex city features, and significantly improves generation efficiency and accuracy.

2.5 Urban Feature Extraction and Processing

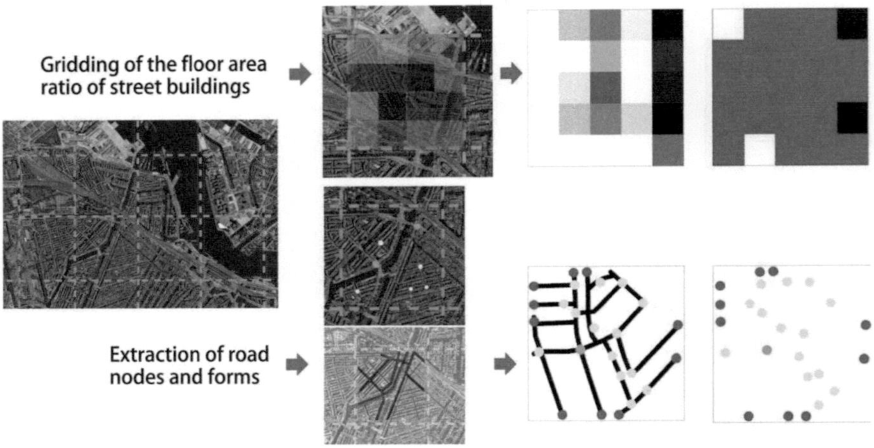

In the data preparation stage, this study extracts building floor area ratio grids and road node networks from satellite imagery as the core input data for the model. First, the urban area is gridded, and the floor area ratio within each grid is calculated to generate a building density distribution map. Second, the road network structure is identified, extracting different types of road nodes and connectivity relationships to form transportation network data.

2.6 Training Model

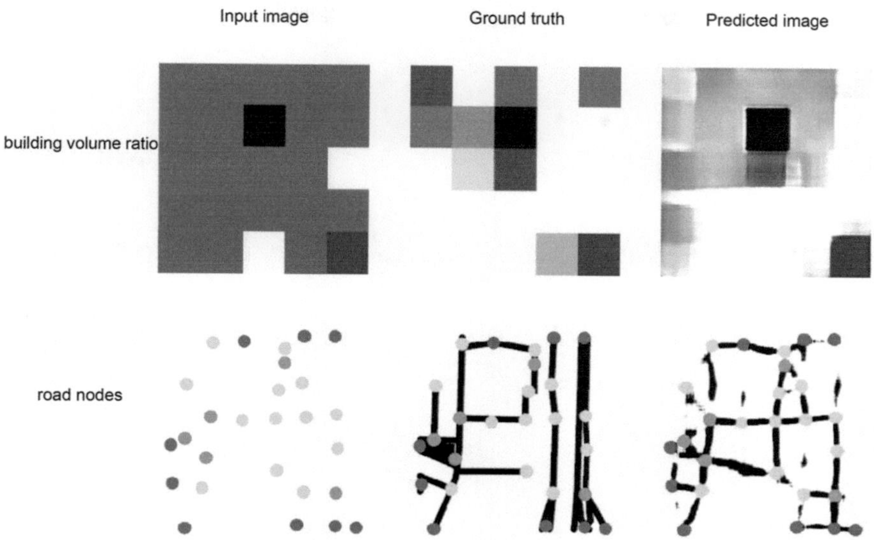

This study trained two independent urban generation models: one for predicting building floor area ratio grids and the other for generating road network structures. The first model, using known floor area ratio data, automatically generates building density distributions for other urban areas, aiding land use planning optimization. The second model, by specifying road nodes, creates a well-connected road network, supporting transportation planning and urban expansion forecasting.

2.7 City Generation

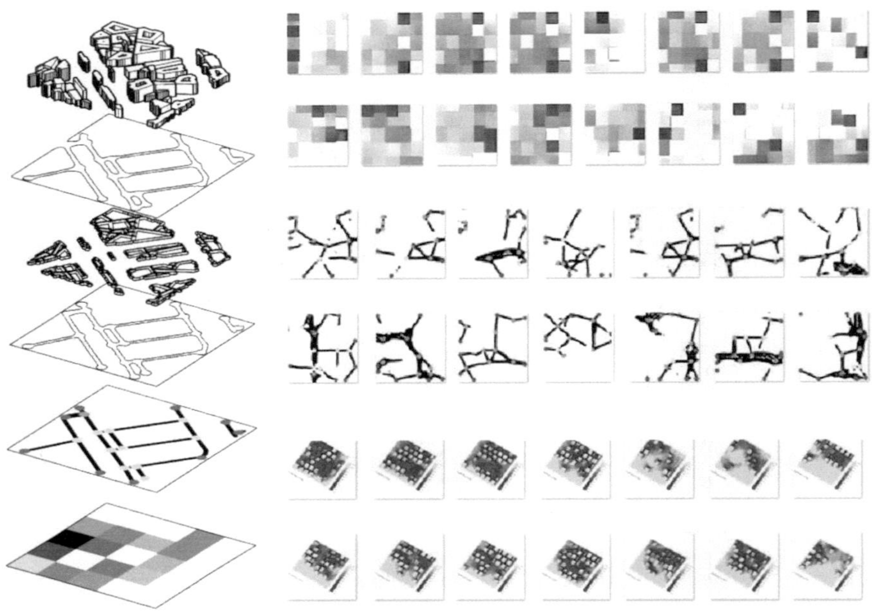

This section outlines a deep learning-based urban generation process, combining building floor area ratio grids and road network structures to form a complete urban spatial form. First, the road network defines building areas, creating rational plot divisions. Then, using the generated floor area ratio grid, building volumes are calculated based on density distributions and extruded into 3D urban forms. The road structure ensures spatial connectivity, while the floor area ratio data determines building height and volume. This method automates the transformation from road networks to building volumes, providing a data-driven, structurally rational, and scalable optimization approach for urban planning.

2.8 Optimization of Results

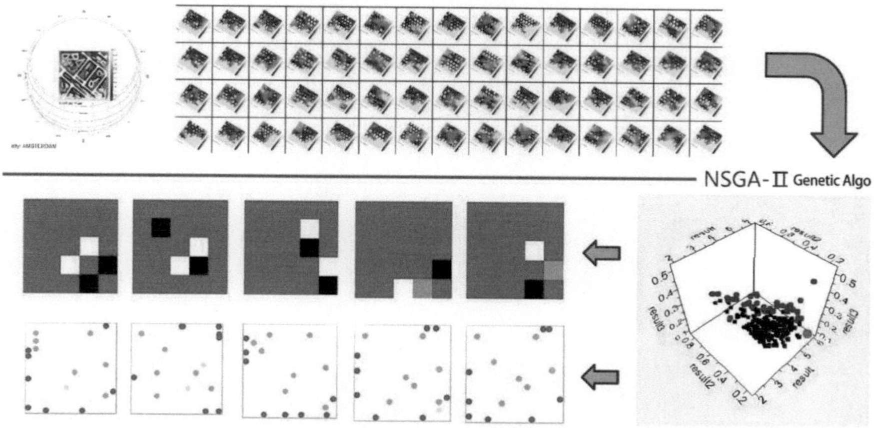

This image illustrates the optimization process of generated urban layouts for solar performance, using the NSGA-II genetic algorithm for multi-objective optimization to balance building layout, road structure, and solar exposure conditions. The top section presents initially generated urban layouts with their corresponding solar radiation analysis, where different colors represent varying solar exposure durations (e.g., red areas indicate good sunlight, while blue areas indicate poor exposure). The right-side 3D scatter plot visualizes the optimization process, where black and blue dots represent different optimized solutions. Finally, the bottom left section presents the optimized results, where the refined building volume grids (red areas) and road networks (colored nodes) have been adjusted to improve sunlight exposure while maintaining urban connectivity. This method integrates deep learning for urban form generation with genetic algorithm-based solar performance optimization, ensuring a more intelligent and sustainable urban planning approach.

Chapter 3
Architectural Metamorphosis

Morphological Optimization of Environmentally Adaptive BIPV Modular Buildings

Members
Guo Mingyue, Jiang Qianli, Zhang Yulin, Yan Lina, Dong Yuxiang, Ji Hongfei

Background
To reach carbon neutrality by 2050, China still faces high carbon emissions. To maintain comfort and GDP growth, we need to use renewable energy sources to reduce building emissions. Building Integrated Photovoltaics (BIPV) combines user needs with energy production, cutting reliance on fossil fuels. Studies suggest BIPV can achieve up to 70% energy savings. With advancing technology and lower costs, BIPV is promising. Modular construction offers benefits like quick assembly and cost-effectiveness. This study proposes creating an adaptive BIPV modular construction optimization tool.

Method
1. Building Form Generation Model

This model utilizes four decision variables: core tube control, building block cantilever height control, building block base height control, and plot cutting control. By optimizing algorithms for building blocks, the model generates building forms entirely controlled by parameters, using a 4848 m base and 33*3 m basic modules.

2. Environmental Performance Model

Building models undergo simulations for three indicators: general thermal climate index, wind uniformity, and compliance rate of comfortable wind speed, generating multiple datasets. Using form and performance parameters, machine learning algorithms are trained to obtain an automated model for specified form environmental performance indicators, selecting regression models for comparison and further optimization with Wallacei X genetic algorithm.

J. Yao, *Intelligent Exploration in Sustainable Design*,
https://doi.org/10.1007/978-981-96-6743-7_3

3. Energy Consumption Model

An energy consumption prediction model is developed using environmental parameter indicators and fossil energy consumption as inputs. Parametric information about the building—such as surface area, usage area, form factor, and volume—is inputted to output fossil energy consumption, which is optimized using genetic algorithms to find the optimal solution.

Significance
This model links building parameters with fossil energy consumption to optimize energy use and design suitable building forms for specific sites. By using standardized modular scales for construction and adding photovoltaic panels on facades, it promotes economic and ecological benefits by reducing energy consumption and facilitating construction.

3.1 Technology Method

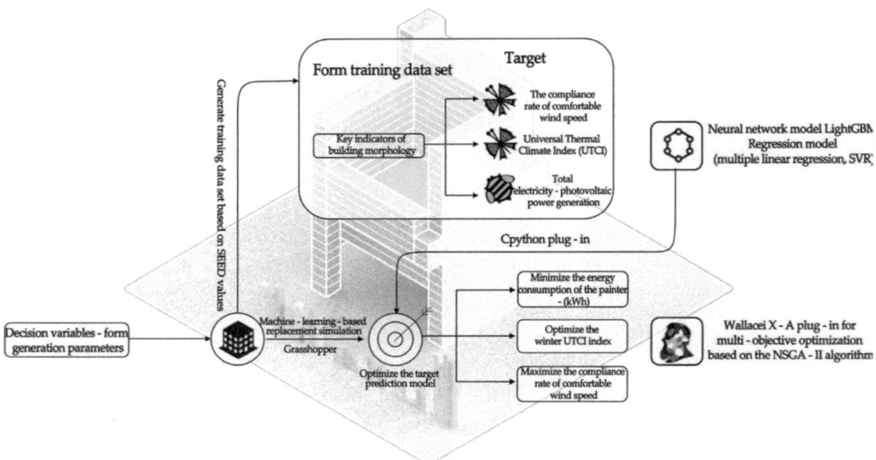

This technical framework integrates machine learning, regression models, and NSGA-II multi-objective optimization to achieve intelligent building form optimization. First, Grasshopper parametric modeling is used to generate various building forms, extracting key performance indicators such as wind comfort compliance, Universal Thermal Climate Index (UTCI), and photovoltaic power generation, forming a training dataset. Then, LightGBM, linear regression, and support vector regression (SVR) models are trained to establish the relationship between building morphology and performance indicators, accelerating the prediction process. Next, the Cpython plugin is utilized for optimization, with objectives including minimizing energy consumption, optimizing winter UTCI, and improving wind comfort compliance. Finally, Wallacei X, combined with the NSGA-II genetic algorithm, performs multi-objective optimization to select the optimal building form balancing comfort, energy efficiency, and sustainability.

3.2 Form Generation

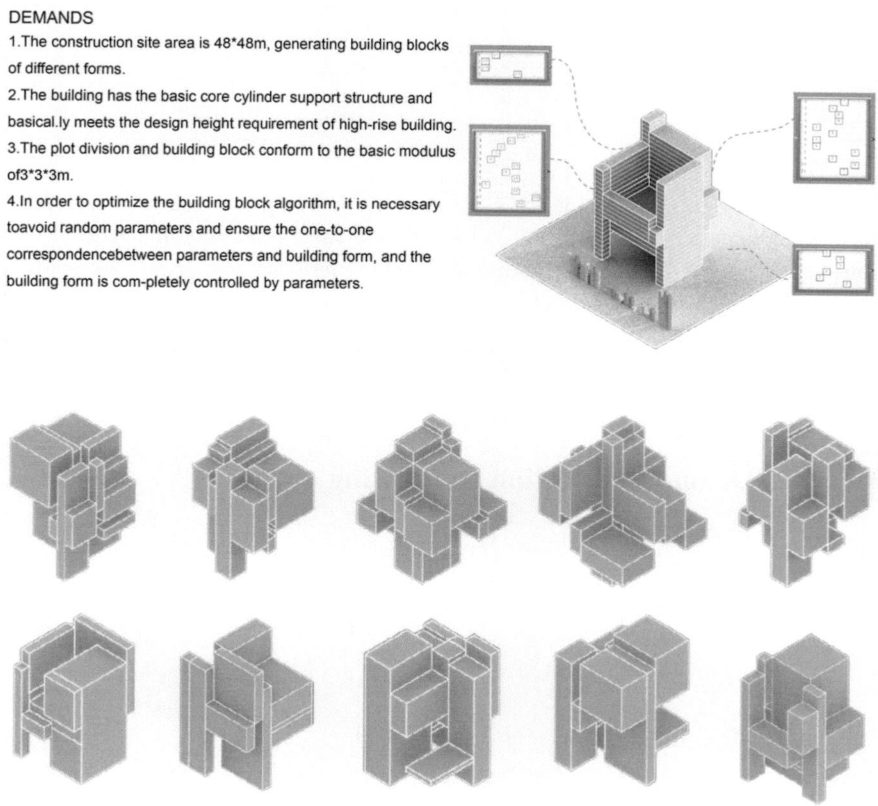

This study integrates parametric modeling with modular design, using a 3 m × 3 m × 3 m standard grid to divide building blocks, ensuring controlled and flexible form generation. All building schemes are parameter-driven, avoiding randomness in design and ensuring adaptability for further optimization. The building adopts a core cylinder support structure, meeting the stability requirements of high-rise buildings while exploring various spatial configurations through different block combinations.

3.3 Energy Simulation Operational Logic

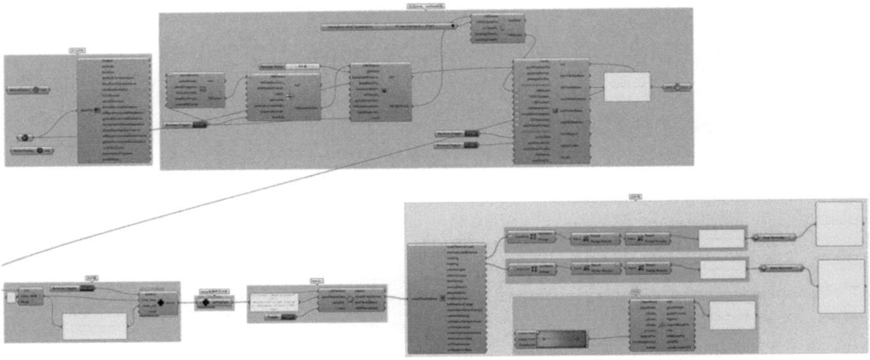

This study utilizes Grasshopper as a parametric modeling platform and integrates Ladybug, Honeybee, and OpenStudio for building energy simulation and optimization. Ladybug provides climate data inputs, including solar radiation, temperature, and wind speed, while Honeybee is responsible for constructing the energy model and calling OpenStudio as the simulation engine, which runs EnergyPlus for thermal environment and energy performance analysis.

3.4 PV Parameter Selection and Setting

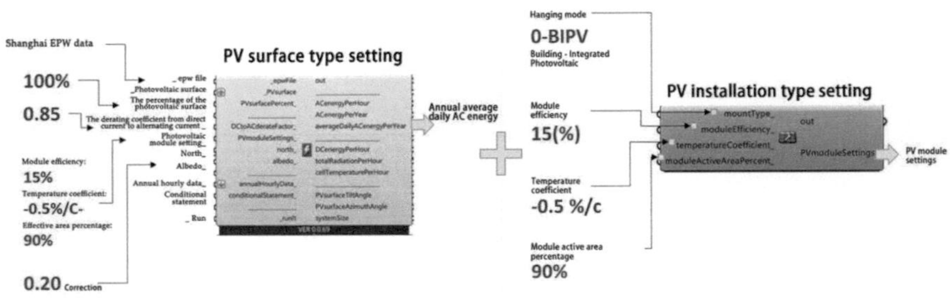

Based on EPW climate data and parametric modeling, the study sets PV surface types and installation methods to optimize building photovoltaic power generation potential. First, the PV surface type setting calculates the power generation capacity of different building surfaces (such as roofs and facades) for photovoltaic installation.

Then, in the PV installation type setting, a BIPV (Building-Integrated Photovoltaic) system is configured, adjusting parameters such as module efficiency, temperature coefficient, and active area percentage.

3.5 Wind Environment Computing Logic

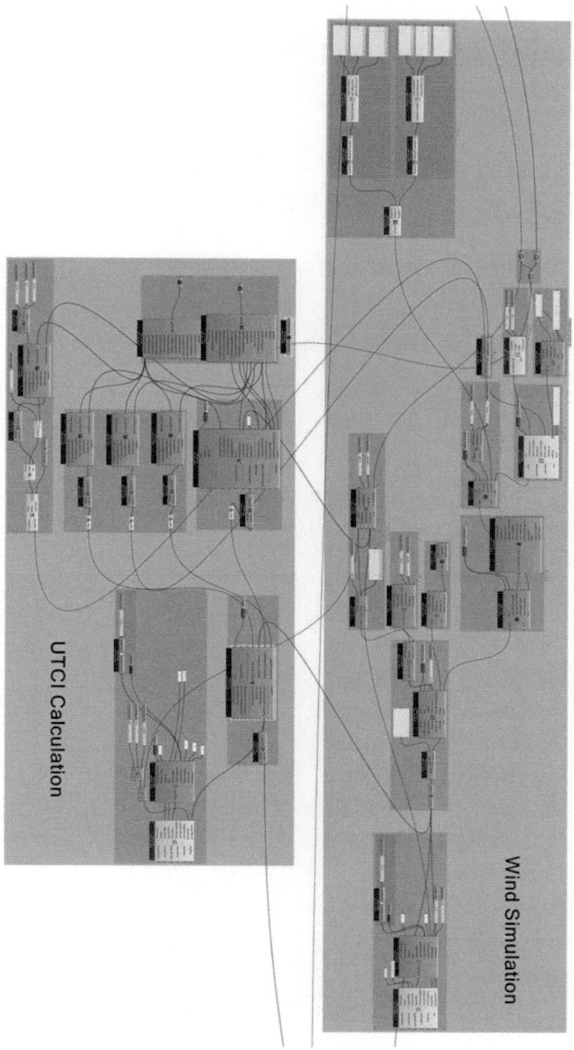

This section illustrates the wind environment computing logic, including UTCI (Universal Thermal Climate Index) calculation and wind environment simulation. The left part calculates UTCI, which evaluates the thermal comfort around the building based on input climate data. The right part of the simulation analyzes the impact of building forms on wind speed and direction, assessing the building's influence on wind comfort. These calculations are automated through various components in Grasshopper.

3.6 Visualization of Environmental Optimization Indicator

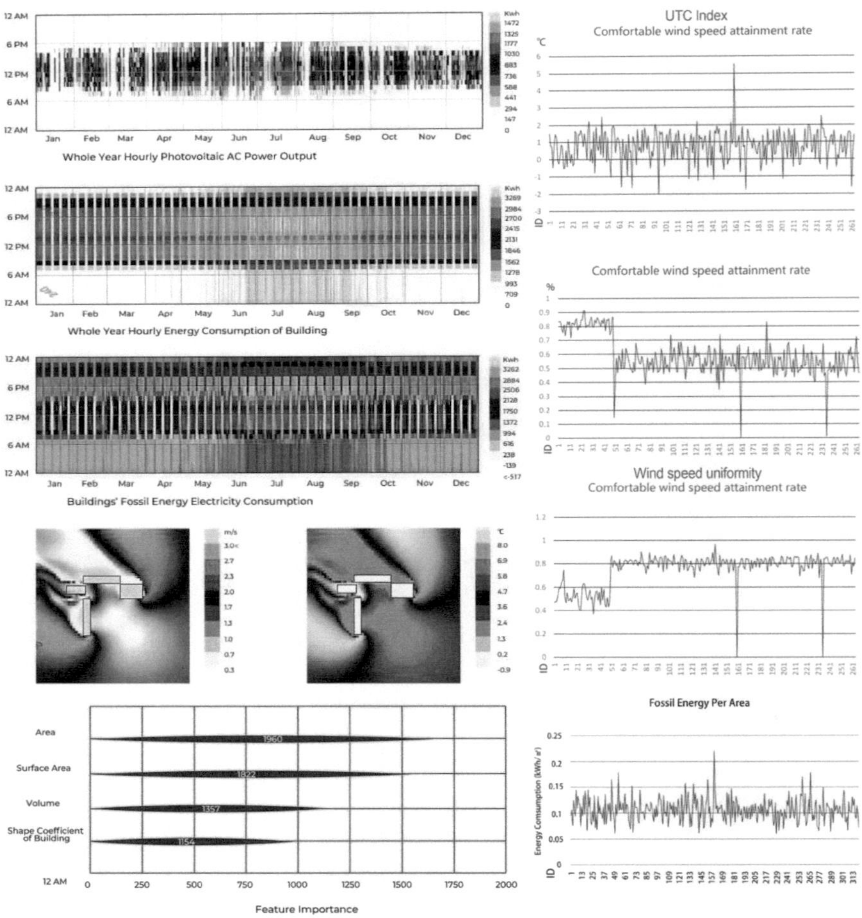

This image visualizes building environmental optimization indicators, focusing on photovoltaic power generation, energy consumption, and wind environment performance. It assesses energy efficiency by showing annual photovoltaic output and energy consumption changes. Analysis of wind speed comfort and uniformity optimizes the wind environment. Additionally, feature importance analysis highlights the impact of building parameters (e. g., area, volume, shape coefficient) on energy consumption, aiding design optimization, energy reduction, and comfort improvement.

3.7 Optimization Process

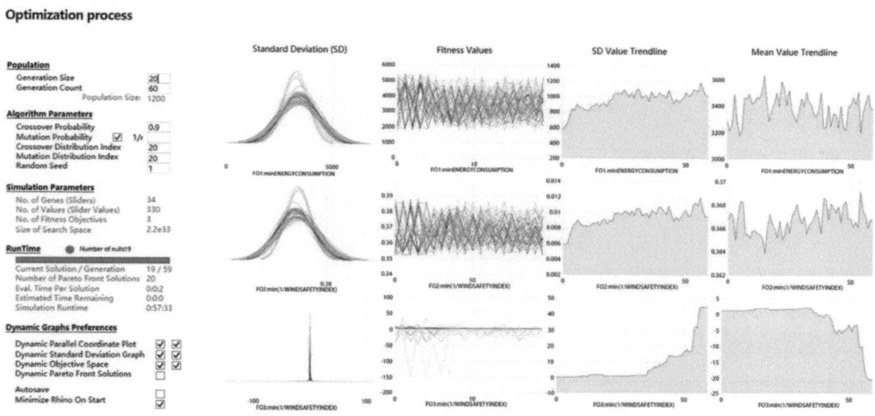

This image illustrates the visualization of the optimization process, focusing on multi-objective optimization to improve building performance, including energy consumption and wind safety. It shows the algorithm and simulation parameters, such as crossover probability and mutation probability, which influence the evolution of the population during optimization. The results display the standard deviation (SD), fitness values, and trends of mean and SD values throughout the optimization, reflecting the stability and progress of the process. The Pareto front solutions represent the balance between multiple objectives. This optimization process, using genetic algorithms, ensures a balance between energy efficiency and wind safety, aiming for the optimal building design.

Chapter 4
Supercompile, Compile, Change

Renewal Planning for Old Town Squatters with Environmental Performance Intervention

Members
Sui Yingda, Shao Yiming, Zhou Zhiwei, Liu Deli, Chen Cuilan, Xu Shiyi

Background
The existing large number of shantytowns in Shanghai suffer from high density, weak lighting, poor ventilation, and other issues. These shantytowns hinder the urbanization process, as the living conditions within them fail to meet residents' needs. When facing urban renewal, the key focus of this research is how to optimize environmental performance and create pleasant public spaces while preserving the existing urban elements as much as possible. Building a new design logic to achieve these goals becomes the focal point of this study.

Method
1. Direction One: Regeneration

The generation is based on dividing the shantytowns from large to small according to the urban texture. Continuous L-shaped divisions are used to delineate alleys, followed by unequal division of each specific cluster, screening and generating these architectural blocks with evenly distributed courtyards or atriums. Multiple generated

J. Yao, *Intelligent Exploration in Sustainable Design*,
https://doi.org/10.1007/978-981-96-6743-7_4

schemes undergo daylight simulation using the Ladybug plugin to obtain a dataset of lighting conditions. Architectural parameters such as building width, depth, alley width, and housing reduction ratio are set, and optimization is performed using genetic algorithms to achieve the optimal balance between plot ratio and average winter daylight exposure.

2. Direction Two: Reconstruction

This approach directly reconstructs a new urban texture by generating algorithms based on the original plot's building density to create node spaces and divide plots. Architectural blocks are generated, and the texture is created based on the original building density. Promising results undergo daylight analysis. Due to the Ladybug plugin's low efficiency, a CNN machine learning algorithm is used to establish a regression algorithm representing 50 architectural features, which is incorporated into the genetic optimization process to obtain the optimal solution set.

Significance

Direction One, Regeneration, seeks a balance between environmental performance, plot ratio, and preservation of the old urban fabric. Direction Two, Reconstruction, emphasizes restructuring urban texture while integrating optimized design with environmental performance parameters. Both approaches offer guidance for future residential area renovations from a technical standpoint.

4.1 Technology Method

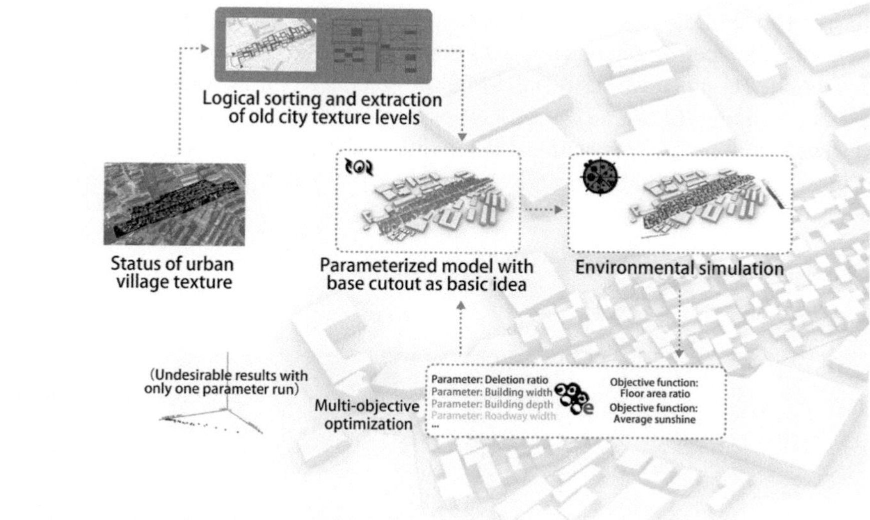

Regeneration: seeking a balance between environmental performance, floor area ratio and preservation of the old urban fabric

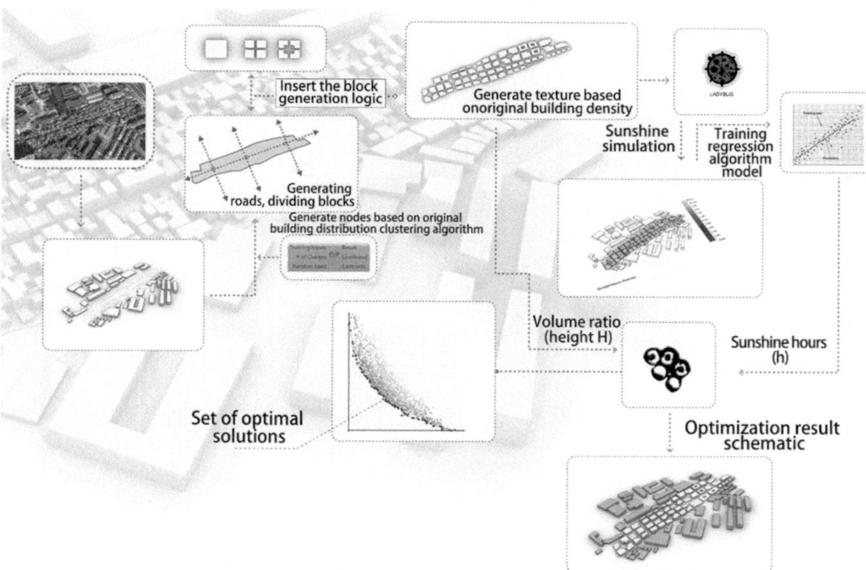

Reconstruction:Reconstruct its urban fabric and optimize the design with environmental performance parameters.

This technical approach combines parametric modeling, environmental simulation, and multi-objective optimization to optimize building design in urban regeneration. The model design parameters include building width, building depth, and road width, adjusted through block generation logic to ensure the rationality and adaptability of the building layout. The regression model, as a surrogate model, predicts environmental performance indicators such as sunlight duration and solar exposure, providing objective functions for multi-objective optimization. Through these optimization algorithms, an optimal set of design solutions is generated, balancing building functionality, environmental benefits, and urban fabric preservation.

4.2 Site Analysis

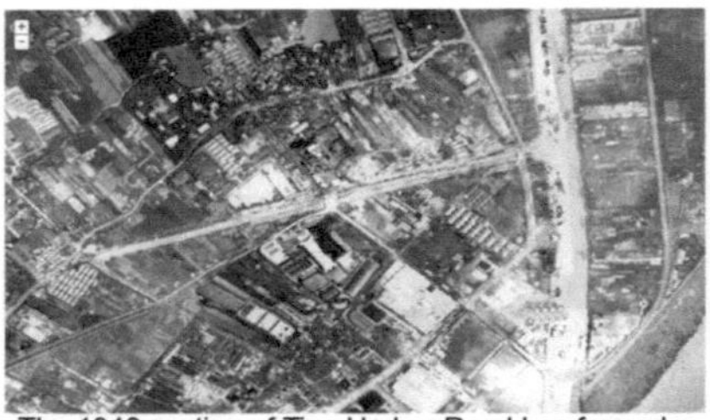

The 1948 section of Ting Harbor Road has formed a wide road surface

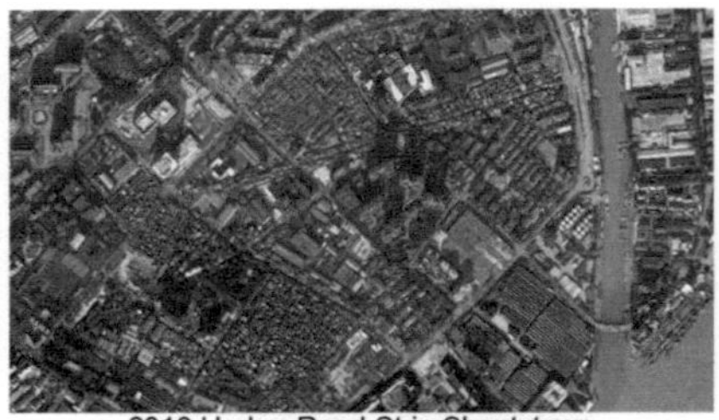

2018 Harbor Road Strip Shantytown

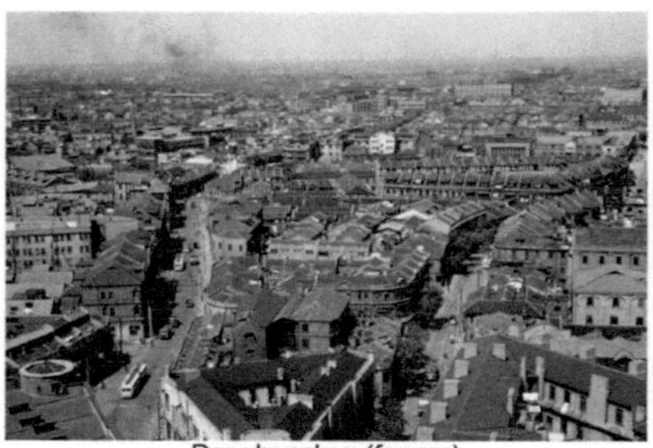

Donghongkou (former)

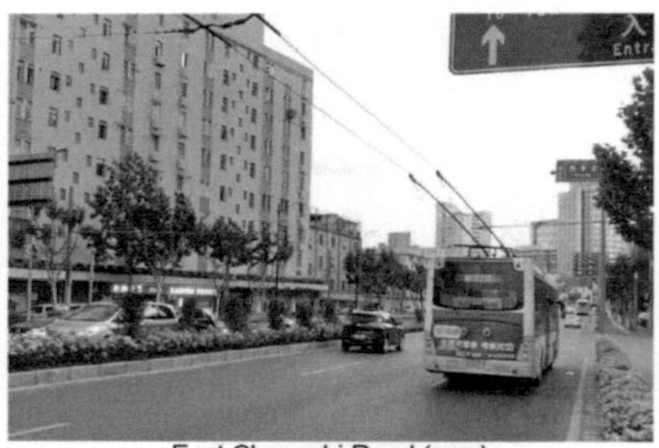

East Changzhi Road (now)

This technical approach combines parametric modeling, environmental simulation, and multi-objective optimization to optimize building design in urban regeneration. The model design parameters include building width, building depth, and road width, adjusted through block generation logic to ensure the rationality and adaptability of the building layout. The regression model, as a surrogate model, predicts environmental performance indicators such as sunlight duration and solar exposure, providing objective functions for multi-objective optimization. Through these optimization algorithms, an optimal set of design solutions is generated, balancing building functionality, environmental benefits, and urban fabric preservation.

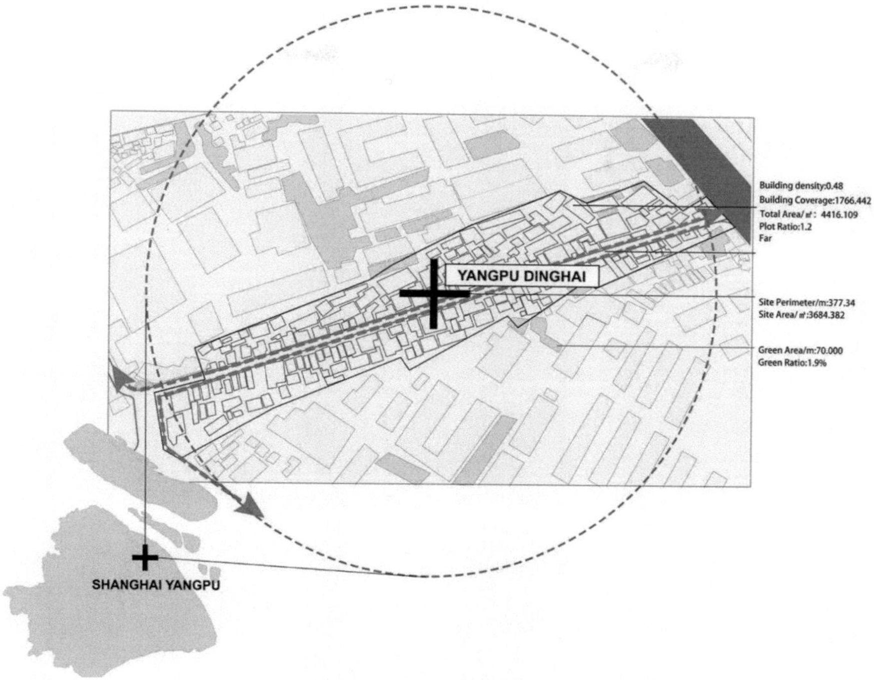

This image presents the site analysis, showing the urban development, building density, and environmental characteristics of the Yangpu Dinghai area from 1948 to 2018 through historical and modern photos and maps.

4.3 Solve Problem

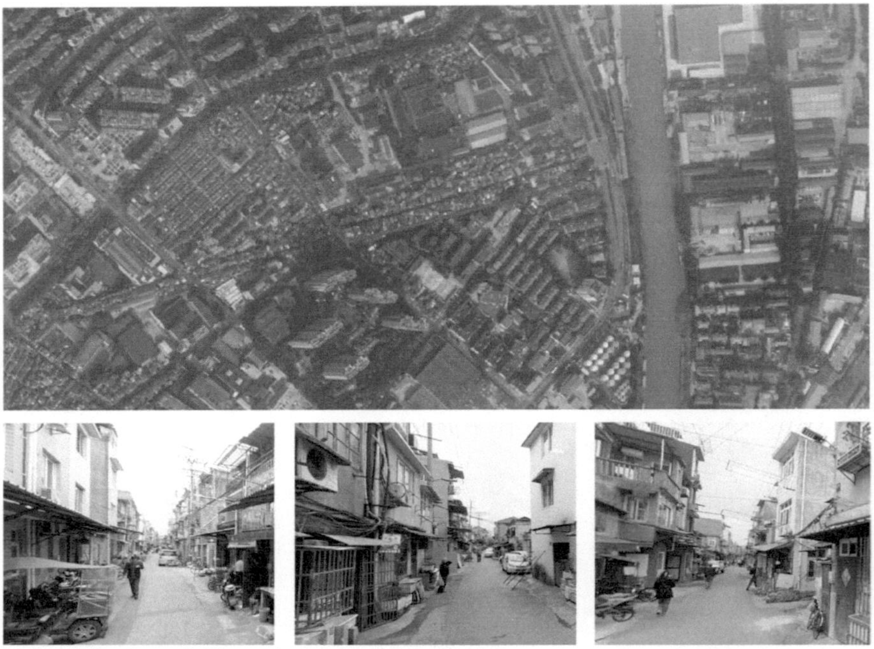

Problem 1: Development of former shanty towns hinders urbanization process.
Problem 2: The living conditions in the original shantytowns are far from meeting the needs of the residents.
Problem 3: How to build a new design logic based on the original urban elements when facing urban renewal.

This image illustrates the issues to be addressed, focusing on three main aspects: the development of the old shantytown, which hinders urbanization, the poor living conditions in the existing shantytown, and the challenge of incorporating original urban elements into the design logic during urban renewal to preserve local characteristics while improving living conditions.

4.4 Design Concept One

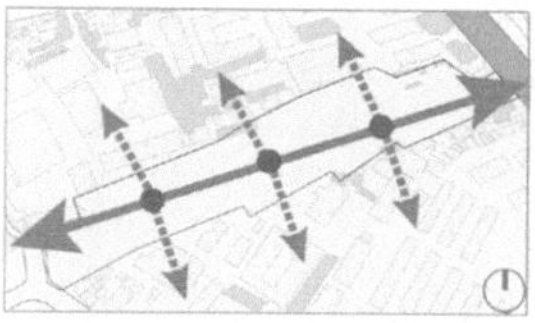

1. Parcel delineation: establishment of the main route and the three main roads of the site

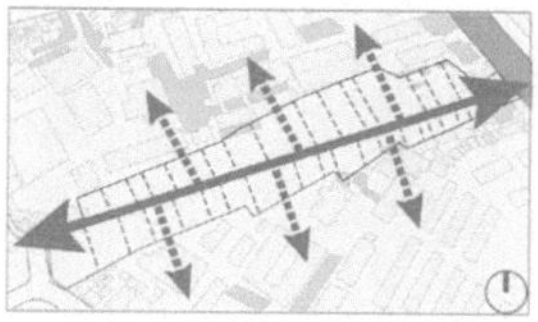

2. Non-isometric (8-24 m) vertical division

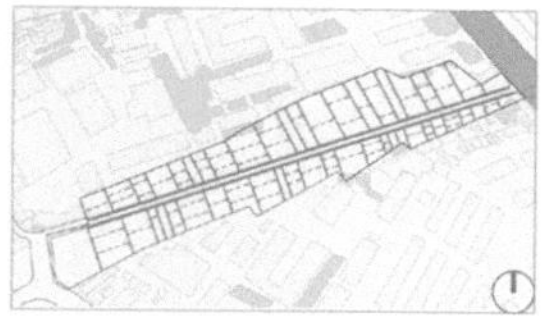

3. Non-equidistant (20-40 m) lateral divisions

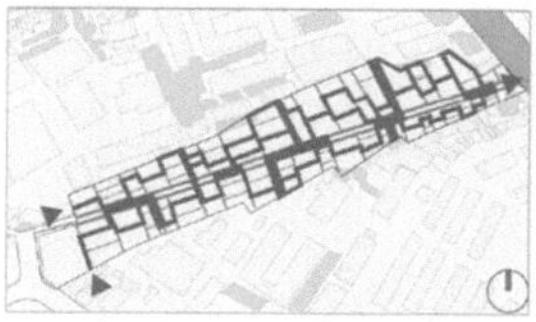

4. Each cluster leaves "L"-shaped lanes along the two edges, making the road network interconnected

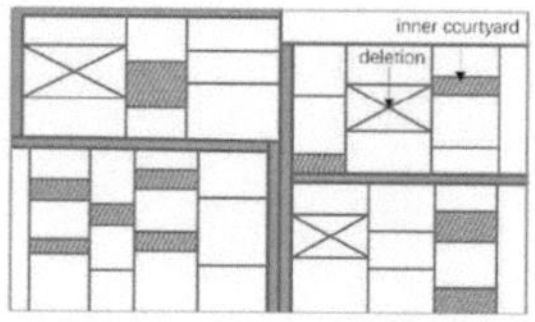

5. Conduct horizontal and vertical unequal spacing division of each cluster, screening smaller area sub-blocks, as the inner courtyard according to a certain index interval deletion as part of the planned demolition, appropriate reduction of density, optimization of environmental performance

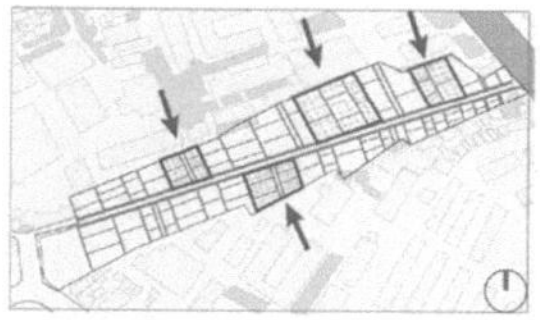

6. Application of the courtyard form to site clusters

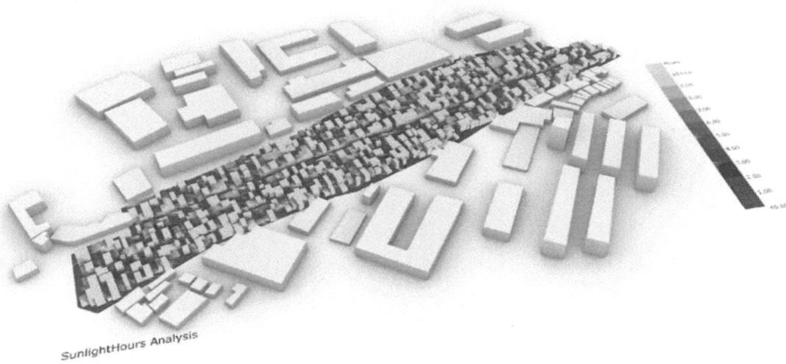

Environment simulation

Design Concept One optimizes the urban form through multiple steps. First, the site is divided into blocks, defining main roads and three primary roads. Then, non-equidistant vertical divisions (8–24 m) and non-equidistant lateral divisions (20–40 m) further partition the area. Each cluster is connected by L-shaped roads, forming an interconnected road network. Next, unequal horizontal and vertical spacing creates smaller sub-blocks, with inner courtyards planned to optimize space density and environmental performance. Finally, sunlight analysis refines the layout, ensuring maximum sunlight distribution and improving space comfort and energy efficiency.

4.5 Design Concept Two

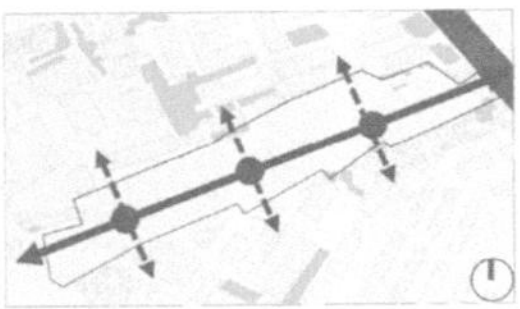

1. Zoning: generating road zones

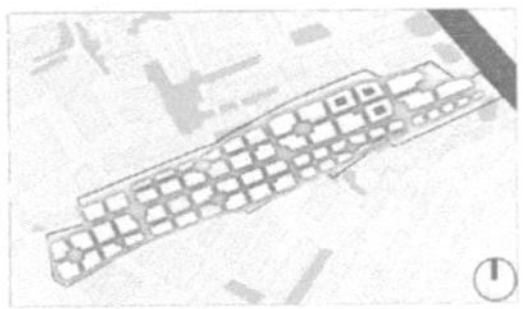

2. Generation of texture based on original density

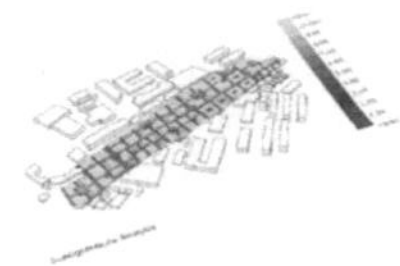

3. Intervention of sunshine simulation in search of excellence

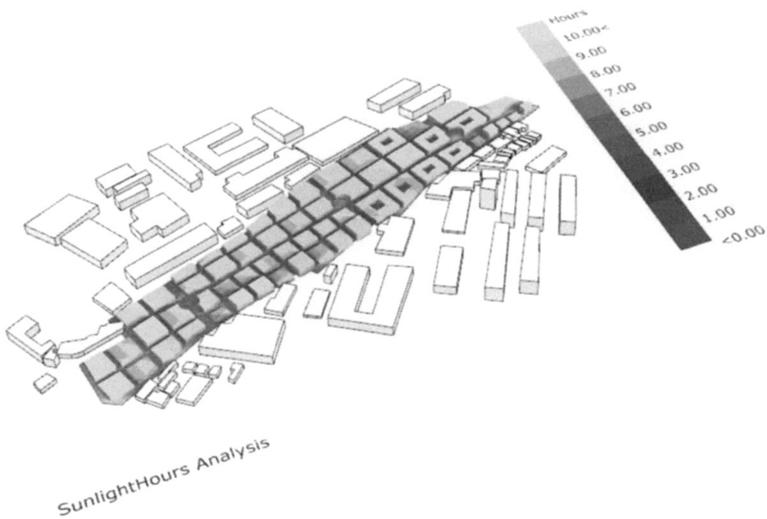

Environment simulation

Design Concept Two adopts a simpler approach, first defining main roads and then dividing the area using non-equidistant vertical divisions (8–24 m) and non-equidistant lateral divisions (20–40 m). It emphasizes clear zoning and a structured layout, avoiding complex inner courtyard designs. Through sunlight duration analysis, it optimizes sunlight distribution for maximum exposure and improved environmental performance. This straightforward approach enhances building sunlight reception and energy efficiency while ensuring functional urban spaces.

Chapter 5
Campus Complex Passive Design Performance-Based Active Search for Excellence

Members
Xu Ke, Li Bin, Xia Tianxiang, He Pan, Lin Duobin, You Baihui

Background
In an urbanized setting, where high-density environments prevail, this study centers on a campus within a typical urban area. It aims to understand how architectural design, microclimate prediction, and surrounding environmental changes interact. By optimizing building clusters around the central square, it seeks to adapt to the hot and humid climate of the Lingnan region, ensuring comfortable campus conditions. Additionally, UTCI will be used to enhance the existing landscape.

Method
1. Basic Model Generation

Three pathways—building orientation, floor plan morphology, and courtyard greening—created independent variables for the building complex. It consists of 5 blocks, each defined by randomly generated orientation and floor plan, totaling 10 factors. With central courtyard greening, there are 11 independent variables. Using an 8×8 m grid module, the overall form is controlled by plan morphology to create controllable blocks with specified height and stories. Additionally, 5 building volumes, including corridors and top blocks, complete the model.

2. Analysis and Optimization

Dependent variable data stem from three analyses: visibility, radiation, and UTCI. Visibility analysis considers human viewpoint height and a 150 m viewing range

J. Yao, *Intelligent Exploration in Sustainable Design*,
https://doi.org/10.1007/978-981-96-6743-7_5

as key parameters. Radiation analysis selects a typical Guangzhou summer day to minimize site radiation, considering both buildings and green spaces. UTCI analysis sets the dependent variable's minimum value to approach the suitable temperature range, considering observed temperatures exceeding 30 °C.

3. Fitting Learning

Wallacei X optimizes 11 independent variables (5 building morphology, 5 building courtyard position, and 1 green space distribution) and 3 dependent variables (minimum visibility obstruction rate, minimum site average radiation, minimum site average UTCI). Active optimization achieves predictive inference. After 2500 calculations, 120 points are generated for subsequent Pix2Pix training set learning.

Significance

Applying workshop-taught skills, we aim to optimize passive design for campus complexes, fostering proactive improvement in building performance. This integrated approach strives to positively impact research on performance-based architectural design.

5.1 Technology Method

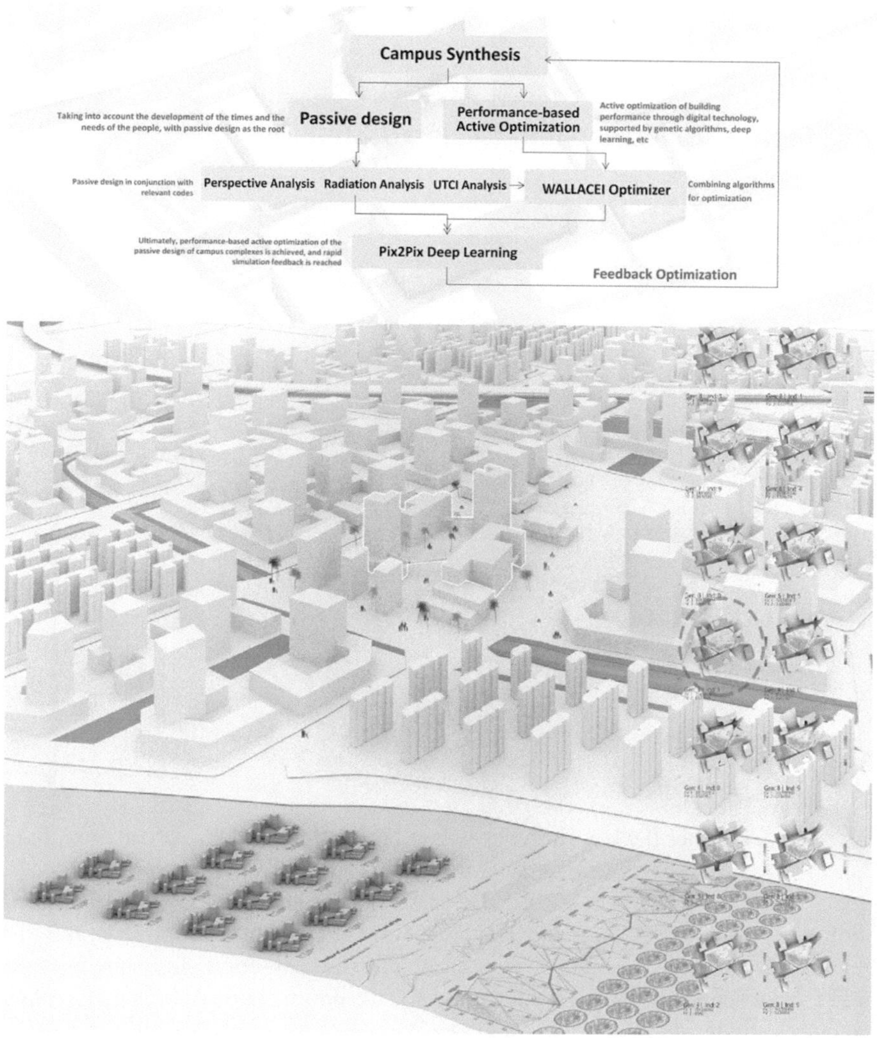

This technical approach showcases a performance-based architectural design process integrating passive design and performance-based active optimization. First, the overall layout is determined through campus synthesis, with passive design as the core to ensure adaptability to evolving needs. Passive design includes perspective analysis, radiation analysis, and UTCI analysis for basic environmental performance evaluation.

Next, the WALLACEI optimizer, incorporating genetic algorithms and deep learning, actively optimizes building performance for optimal outcomes. Finally, the Pix2Pix deep learning model and feedback optimization continuously refine the design, achieving the best building-environment adaptation.

5.2 Center Courtyard and Enclosed Buildings

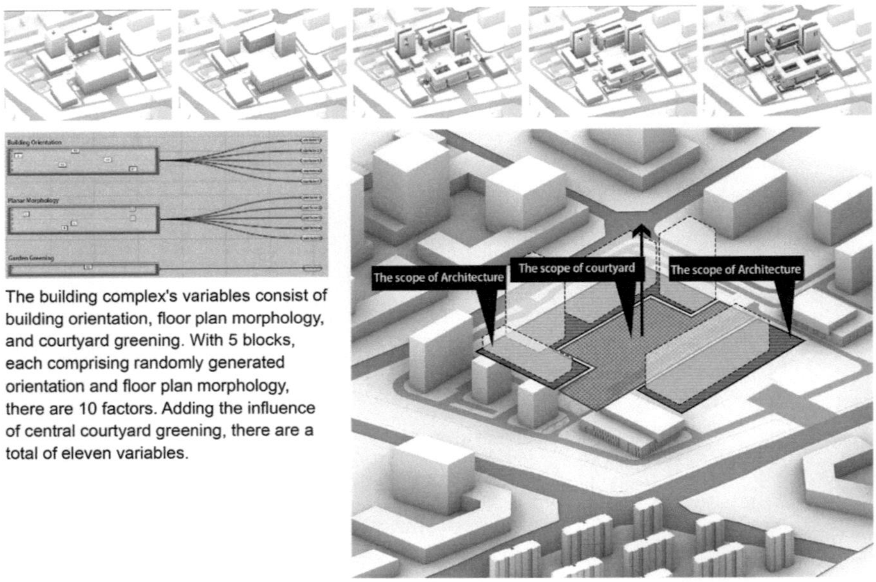

This section outlines the key variables in building design: building orientation, floor plan morphology, and central courtyard greening. The complex includes 5 blocks, each with randomly generated orientations and floor plans, involving 10 factors. Adding central courtyard greening introduces an 11th variable for optimization. The diagram illustrates the distinct scopes of architecture and courtyard, showing how these factors shape the overall layout and guide the optimization of both architecture and courtyard design.

5.3 Architecture Generation Logic

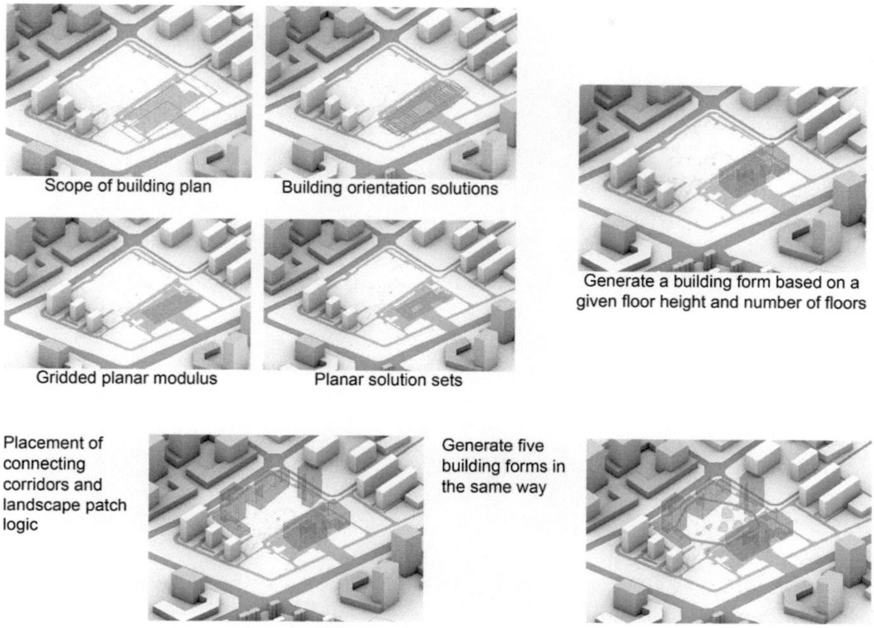

This section outlines the architecture generation logic. First, the building plan scope is defined, setting the development area. Next, various building orientation solutions are generated to align buildings with the site. A gridded planar modulus organizes the layout systematically, followed by creating multiple plan solution sets. Based on floor height and number of floors, a building form is developed to meet design needs. The process includes adding connecting corridors and landscape patches for circulation and aesthetics. Finally, this logic generates five building forms consistently, ensuring a cohesive architectural layout.

5.4 Evaluation Mode

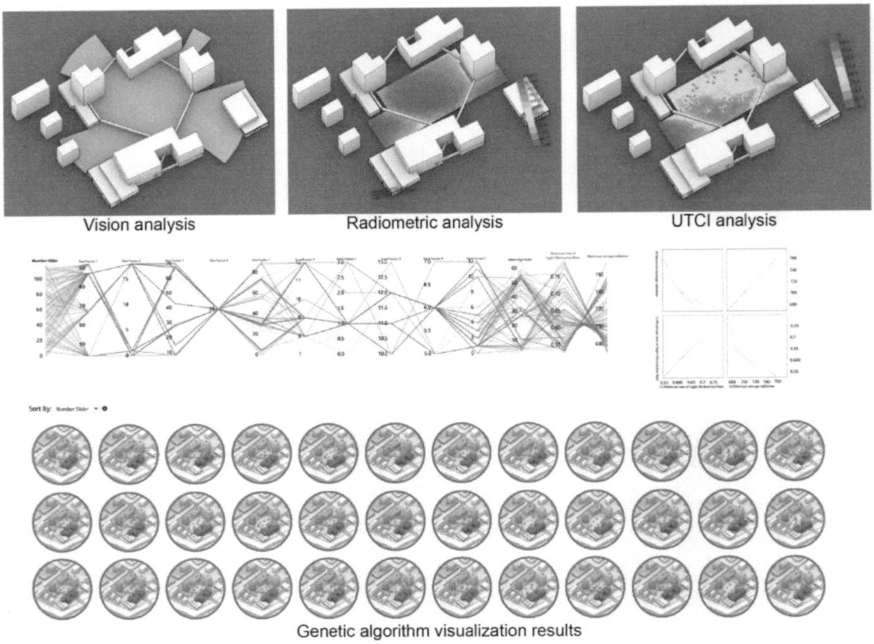

This section introduces the evaluation model for architectural design, assessing building forms through visual analysis, radiometric analysis, and UTCI (Universal Thermal Climate Index) analysis. Genetic algorithms optimize and select the best solutions based on environmental performance. The study uses Design Explorer to visualize genetic algorithm-optimized solutions, displaying results across various performance metrics and offering designers a clearer, actionable optimization decision-making process.

5.5 Light Simulation

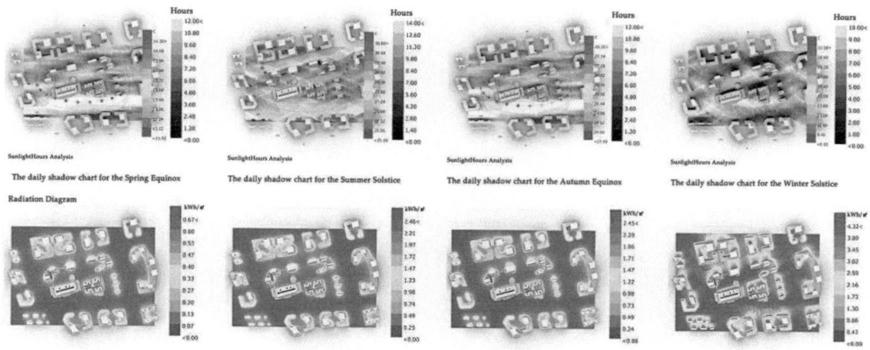

The Ladybug plugin was used for light and radiation analysis. The daily shadow charts for the Spring Equinox, Summer Solstice, Autumn Equinox, and Winter Solstice are displayed.

5.6 Constructing the Training Set Database

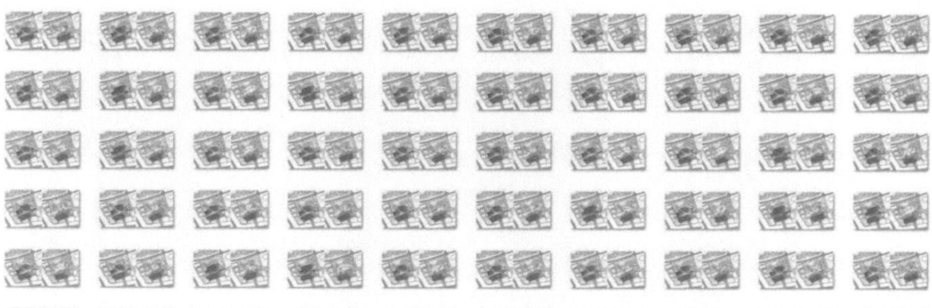

This section explains training dataset construction and machine learning model prediction. Input data on building forms and environmental factors (e.g., sunlight, wind) is collected, and result images are generated via simulations. These input images train deep learning models to predict environmental performance metrics based on provided building forms and environmental data.

5.7 Machine Learning

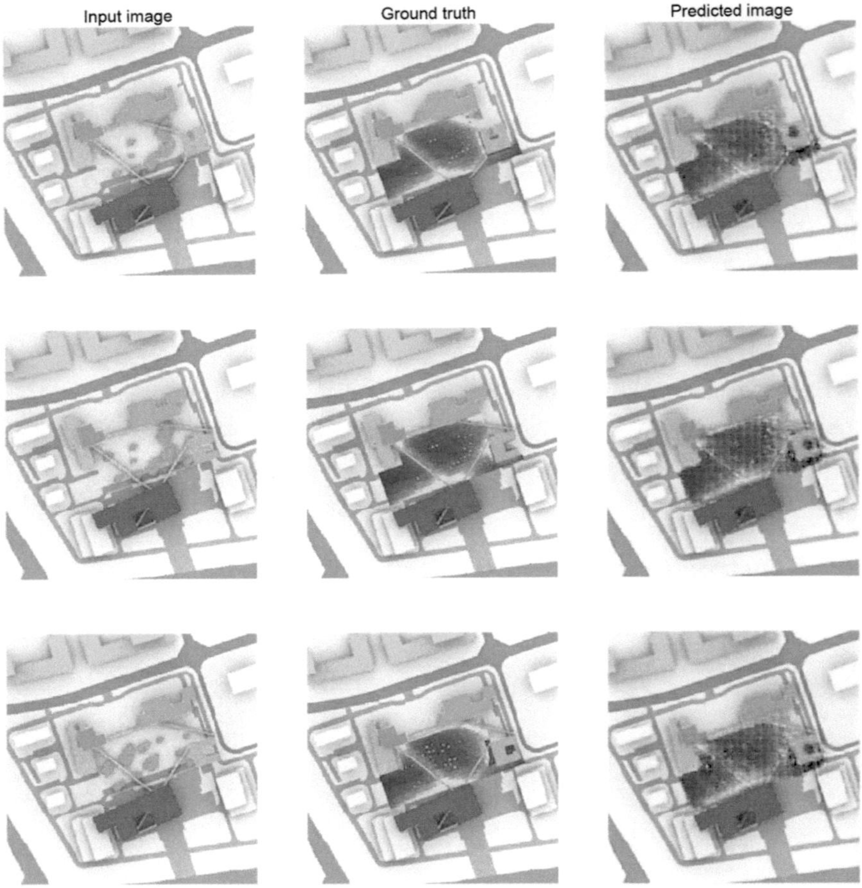

In the prediction phase, input images include building floor plans, surrounding environments, and related data like lighting and temperature, while output images show predicted environmental performance results, such as sunlight distribution. Comparing predicted images with ground truth images evaluates the model's accuracy and effectiveness, offering insights and optimization directions for design.

Chapter 6
Exploring the Method of Evaluating Flooding Risk of Building Groups in Mountainous Area with GAN and GA

Members
Jia Jing, Hu Beiyu, Cui Yuechen, Lu Yijun, Feng Wenhan, Tu Han

Background

With global climate change causing more frequent extreme weather, mountain flash floods present challenges to architecture. Designing buildings in mountainous regions becomes complex, requiring consideration of rainwater runoff and downstream impacts. This study evaluates rainstorm risks for specific building locations, aiming to expedite optimization solutions. It explores using parametric architectural design and neural networks to assess risks in mountainous building designs and seeks optimal solutions.

Method

1. Model Generation

Based on site selection principles considering mountainous terrain, existing villages, and flood risks, diverse case block selection criteria were applied. Sites with existing village clusters and varied terrain, each 350 m × 350 m, were chosen. Block division within site boundaries was guided by area, parcel, and angle constraints, with adjustments using existing building density data.

2. Hydrological Simulation

Instead of traditional methods using QGIS, hydrological simulations were conducted in Grasshopper, studying water droplet trajectories to mimic real-world hydrological patterns. The simulation at 100 frames was found to be most effective.

3. Machine Learning

With 51 sample sets comprising grayscale elevation maps and labeled building blocks, pix2pix machine learning was applied alongside parameter schemes. The

J. Yao, *Intelligent Exploration in Sustainable Design*,
https://doi.org/10.1007/978-981-96-6743-7_6

goal was to expedite optimization solutions by exploring parametric design combined with adversarial neural networks. Results showed this approach is feasible for simulating rainstorm risks in mountainous building design, with building density playing a significant role.

Significance
Single building angles alone aren't sufficient for assessing rainstorm risks; terrain and neighboring buildings must be considered. While current parameters need refining for accuracy and there are inconsistencies in local building layouts affecting sample accuracy, the research partially reveals the link between mountainous architectural layouts and rainstorm risks, establishing a preliminary risk assessment model for mountainous building designs.

6.1 Technology Method

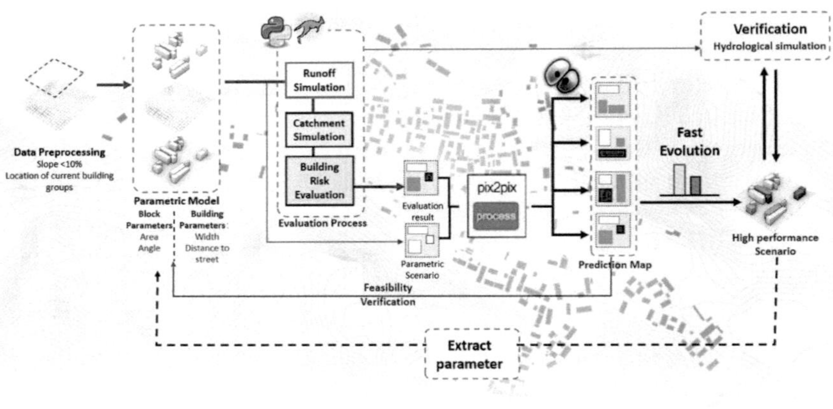

This technical process outlines optimization based on building forms and environmental analysis. First, data preprocessing determines basic block parameters (e.g., building area, angle, width, distance to the street) and analyzes geographical locations and slopes (under 10%) of building clusters.

Next, runoff simulation, catchment simulation, and building risk assessment evaluate scenarios. The Pix2Pix deep learning model then rapidly evolves and predicts to generate the optimal building layout.

Then, prediction results are verified as a high-performance scenario, with hydrological simulations ensuring feasibility. This feedback mechanism optimizes building design to meet environmental performance criteria.

6.2 Case Location

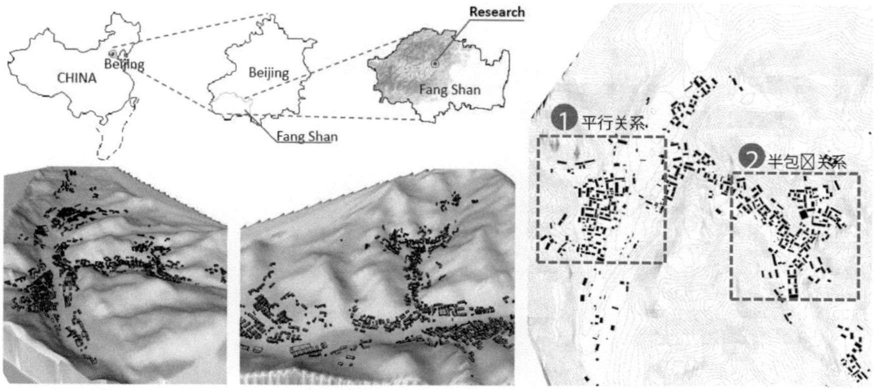

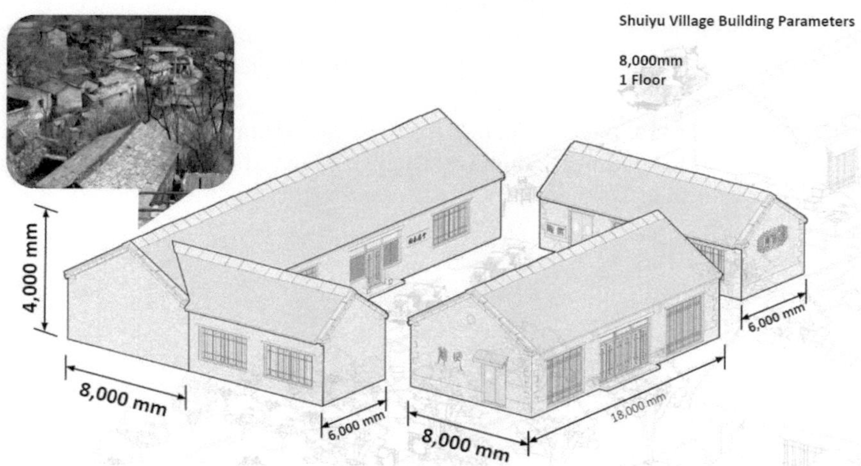

This section presents the case study location and building parameters. The research area is in Fangshan, Beijing, China, focusing on local building distribution. The map highlights the study area and illustrates two building layout relationships: parallel and semi-enclosed. The 3D model shows Shuiyu Village parameters, with 8 m × 8 m single-story buildings, and details each building's dimensions, demonstrating its structure and layout design.

6.3 Research Background

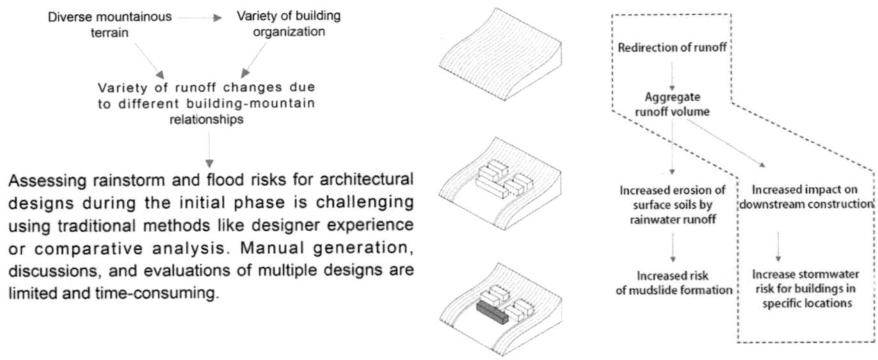

This diagram shows how mountain terrain and building organization affect runoff and flood risks in architectural designs. In areas with complex mountainous terrain and varied building layouts, changes in building runoff can increase risks like soil erosion, landslides, and stormwater hazards. Traditional methods, such as designer experience or comparative analysis, struggle to assess these risks early on, and manual design generation and evaluation are time-consuming and limited. The diagram highlights that increased building runoff raises aggregated runoff volume, worsening soil erosion, mudslide risks, and impacts on downstream construction.

6.4 Parametric Generation

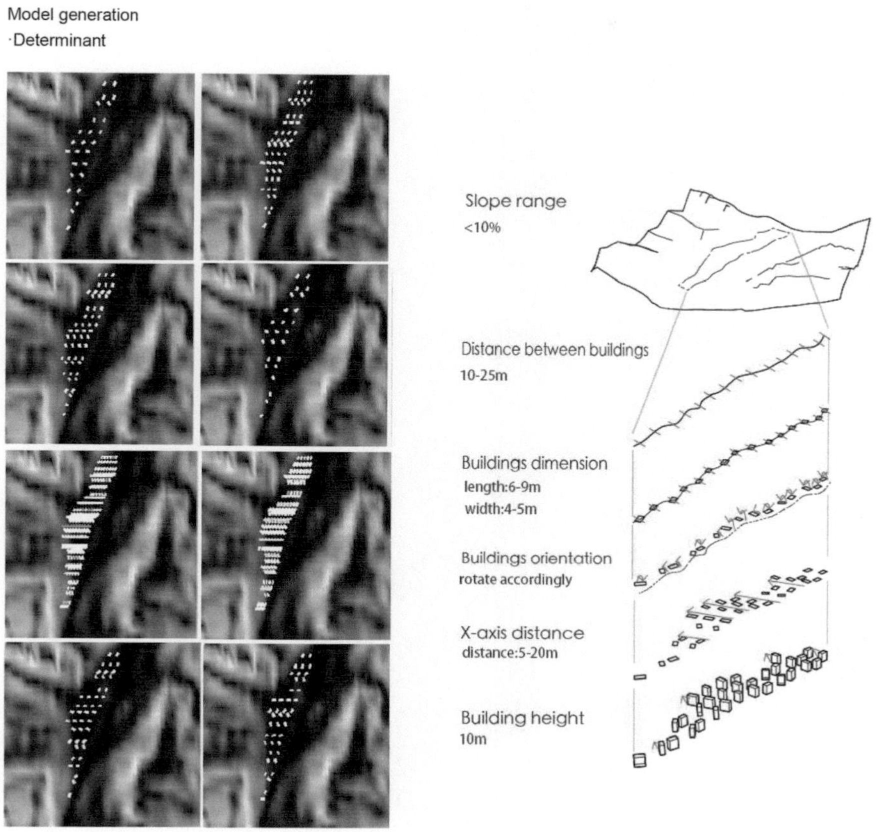

This image shows parametric modeling based on terrain. The building layout and design are controlled by parameters like building size, angle, and distance to the street. By setting slope range (under 10%), building spacing, dimensions (length, width, height), and orientation, various building schemes can be generated.

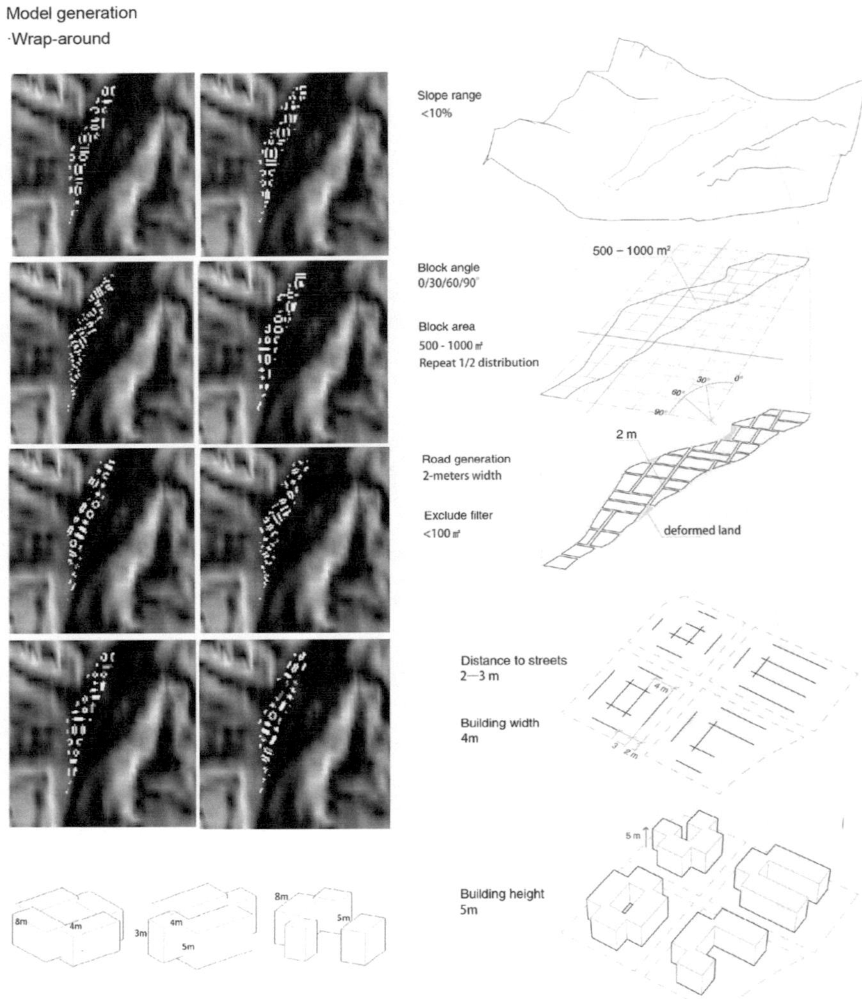

The "wrap-around" model generation method further optimizes the space and generates roads by setting parameters such as the building angle, building area range (500 to 1000 square meters), and road width. The generated road width is 2 m, and the building width is 4 m, ensuring the rationality and feasibility of the urban layout.

6.5 Hydrological Modeling Approach

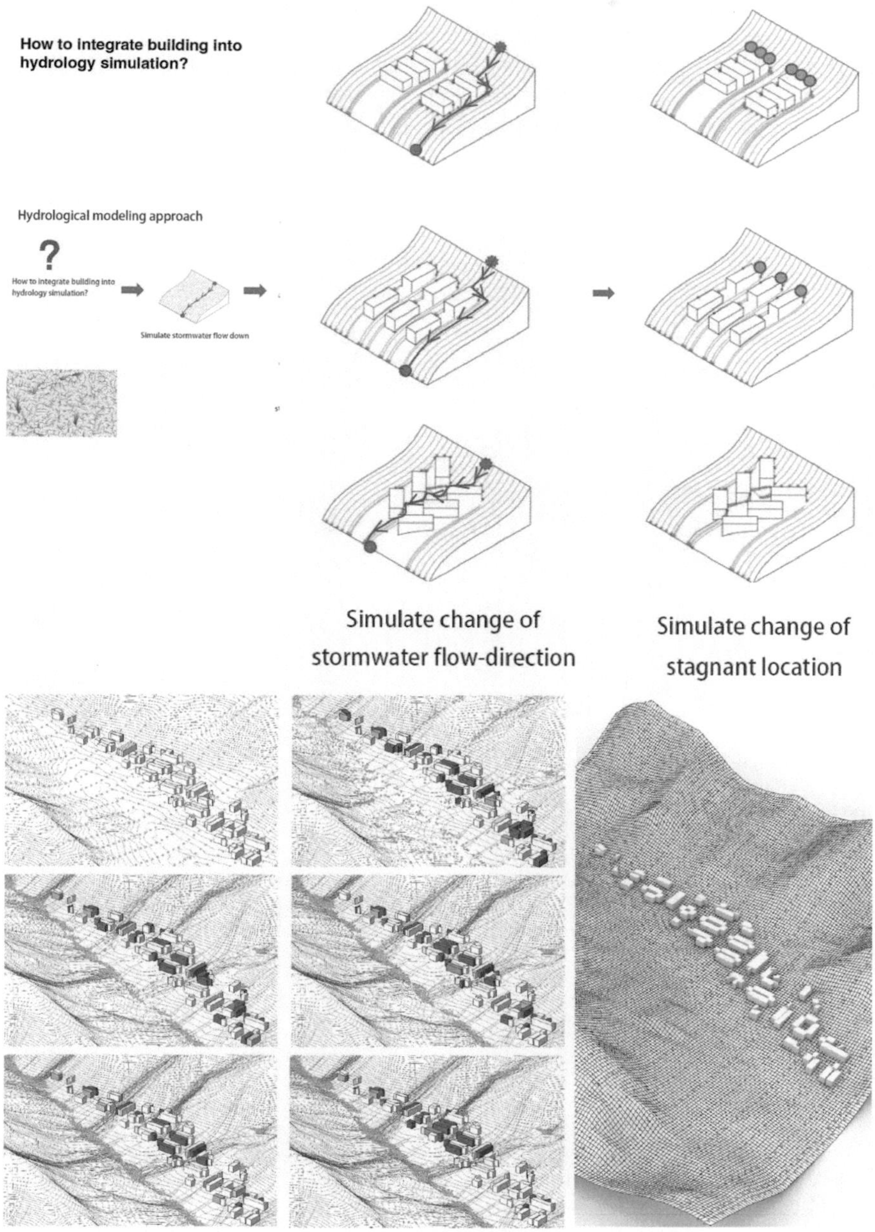

This image demonstrates how buildings are integrated into stormwater simulations to analyze their impact on water flow and stagnation. Using ArcGIS, the runoff simulation first tracks stormwater flow paths to observe how buildings affect water flow direction. It then simulates changes in flow direction caused by buildings and analyzes the impact of different building layouts. Finally, it identifies potential stagnation zones by simulating water accumulation areas around buildings. Each frame illustrates changes in flow paths and stagnation zones with varying building configurations, providing data for hydrological simulations and urban planning.

6.6 Machine Learning

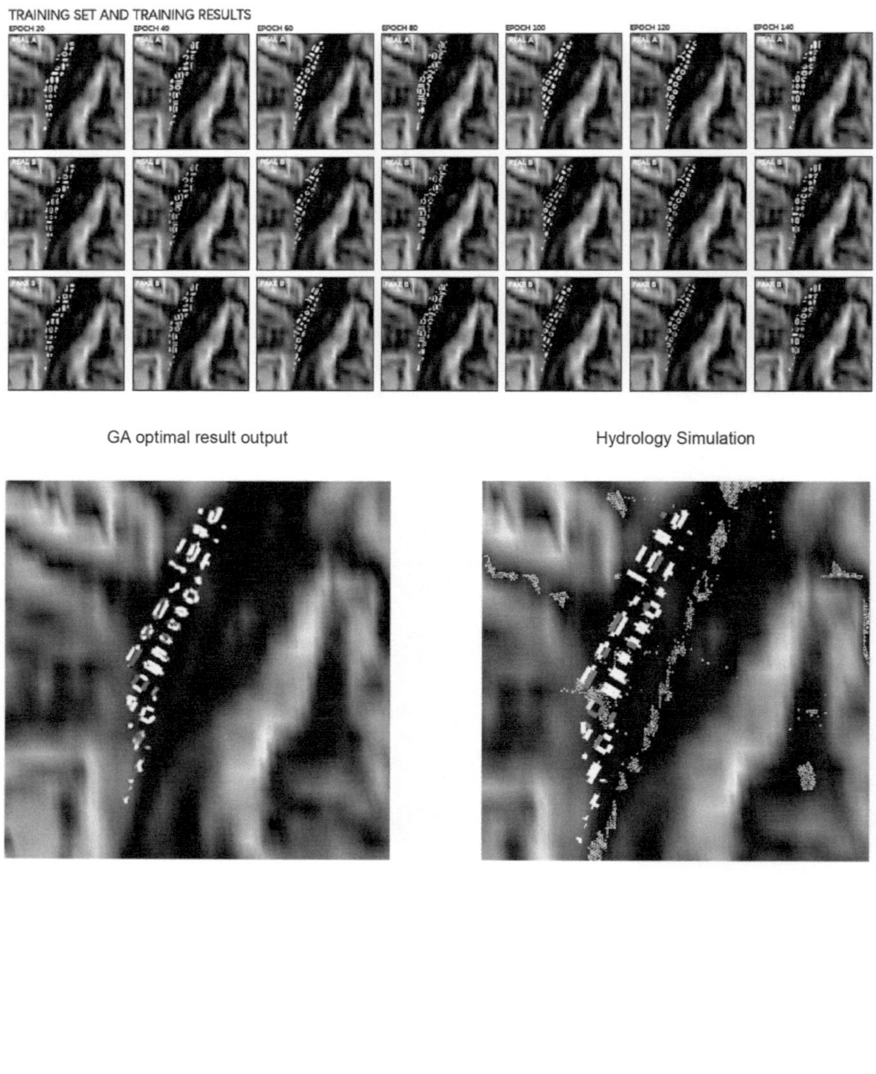

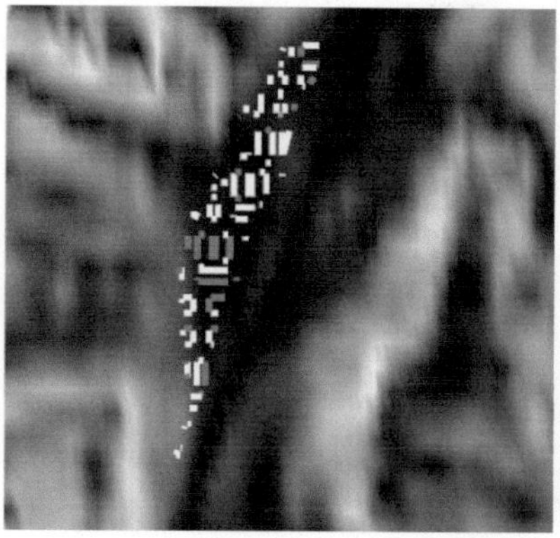

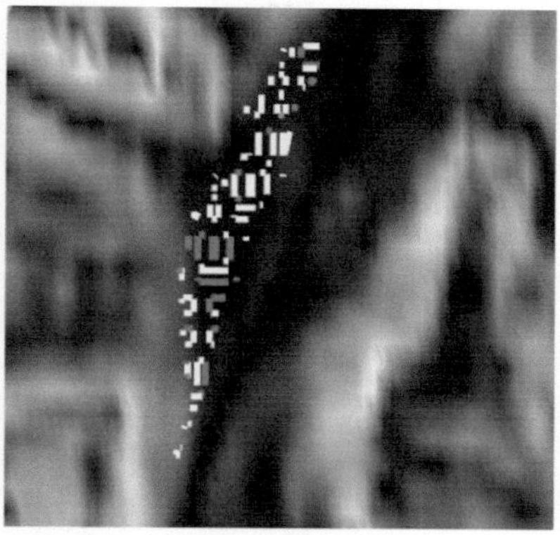

This section shows the Pix2Pix model training to quickly assess stormwater risks for buildings under various terrain conditions. Through multiple training cycles (20–140 epochs), the model improved its accuracy in predicting building layouts and water flow risks. After training, the predictive modeloptimized building layouts using a genetic algorithm (GA), reducing stormwater risks and ensuring better environmental adaptation across different terrains.

Chapter 7
Old City with Wind: Exploring the Design Generation of Public Buildings in Urban Villages

Members
Buyin, Wang Yayu, He Simian, Li Zehao

Background
The site is in an urban village in Baotou, Inner Mongolia, centrally located with diverse businesses and demographics. Urban villages stem from rapid urbanization, encircling rural areas, with unresolved internal contradictions. This study aims to enhance the village environment, focusing on factors like sunlight and wind to improve residents' comfort.

Method
1. Current status of base.

Initially, we conducted on-site research and digital analysis of the surrounding area, covering building characteristics, number of floors, quality, environmental assessment, Points of Interest (POI), landscape gaps, traffic assessment, parking, greenery, and supporting facilities. This informed our design proposal.

2. Initial optimization of design.

We subdivided the land and determined building scales based on surrounding architectural dimensions. Then, we randomly generated building blocks using architectural parameters to form a community. We screened preliminary results by varying parameters such as floor numbers, heights, and orientations. We used the Ladybug

J. Yao, *Intelligent Exploration in Sustainable Design*,
https://doi.org/10.1007/978-981-96-6743-7_7

plugin to optimize sunlight conditions and the Butterfly plugin to simulate wind environments for the best results.

3. Optimization loop.

Lastly, we used a pix2pix adversarial network to iteratively learn from sunlight and wind environment analyses. We then simulated thermal comfort and predicted environmental performance using GAN networks. This evaluation helped optimize generation parameters for refining the architectural layout and its relationship with the environment.

Significance
We simulated experiments on northern urban villages with poor wind conditions. At the neighborhood scale, we established a model for automated generation based on environmental performance optimization, offering suggestions for future urban renewal or neighborhood revitalization projects.

7.1 Technology Method

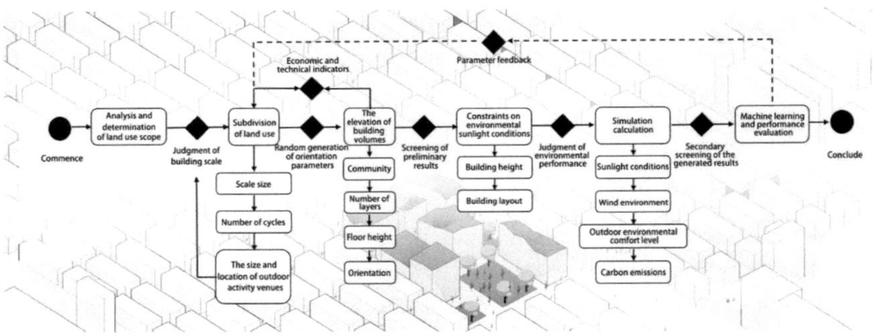

This diagram outlines a city design process optimizing building layouts in multiple steps. Initially, the land use scope is analyzed, followed by determining building scale. Based on this, the land is subdivided, and building orientation and parameters like number of floors, floor height, and layout are randomly generated.

On this basis, environmental sunlight constraints (e.g., building height, layout) are applied, and simulations evaluate sunlight conditions, wind environment, and outdoor comfort. Machine learning then assesses performance to filter optimal designs meeting preset criteria. Finally, feedback optimization refines the design for the best environmental and performance outcomes.

7.2 Basic Site Information

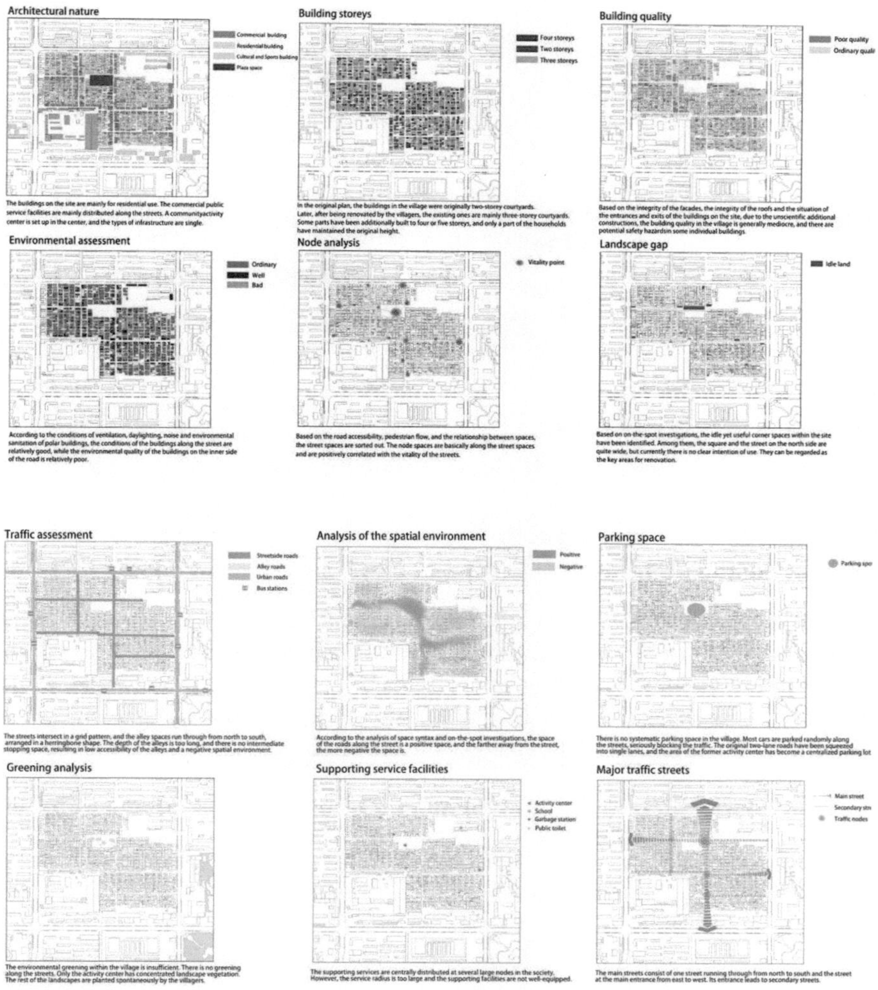

The site analysis covers multiple dimensions. Building type analysis distinguishes commercial, residential, and public buildings; building height analysis shows floor counts. Environmental assessment evaluates quality through sunlight, ventilation, and noise analysis. Node analysis assesses activity points based on road accessibility and walk ability; landscape gap analysis identifies areas lacking green space. Traffic assessment displays street traffic flow, bus stops, and road layouts. Spatial environment analysis reveals positive and negative space relationships, while parking space analysis highlights parking issues. Greenery analysis shows green space distribution and environmental greening. Service facility analysis maps public facilities, and the major traffic street section highlights key traffic routes and nodes.

7.3 Block Generation

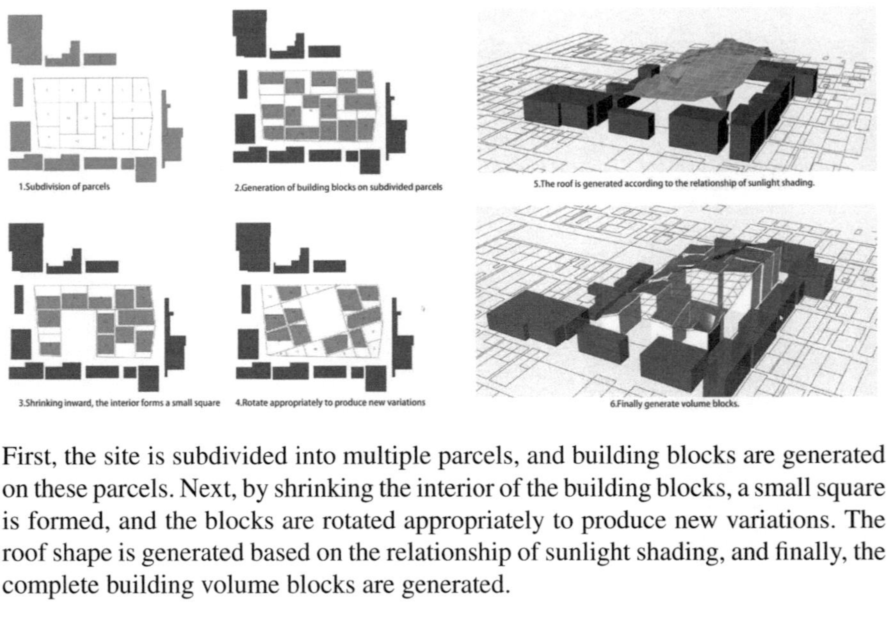

First, the site is subdivided into multiple parcels, and building blocks are generated on these parcels. Next, by shrinking the interior of the building blocks, a small square is formed, and the blocks are rotated appropriately to produce new variations. The roof shape is generated based on the relationship of sunlight shading, and finally, the complete building volume blocks are generated.

7.4 Program Generation

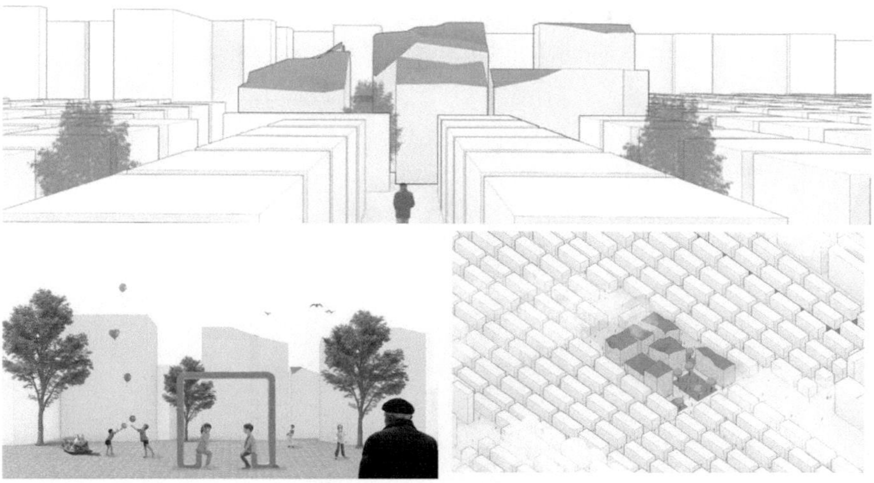

The diagram shows the design of open spaces from different perspectives, including pedestrian streets, green areas, and interactive zones, providing excellent public activity spaces and optimizing the overall accessibility and comfort of the area.

7.5 Daylight Simulation

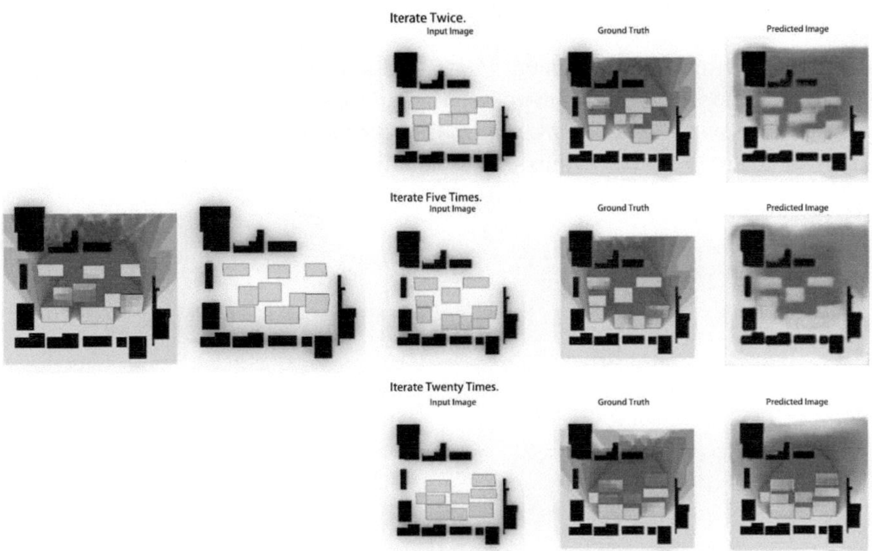

In the daylight simulation, the building layout and sunlight conditions are first simulated to create an initial daylight distribution map. Through multiple design iterations, the trained Pix2Pix model predicts daylight conditions, with predictions compared to actual simulation results. This iterative process continuously optimizes the design, enhancing the building's adaptability to surrounding sunlight.

7.6 Wind Simulation

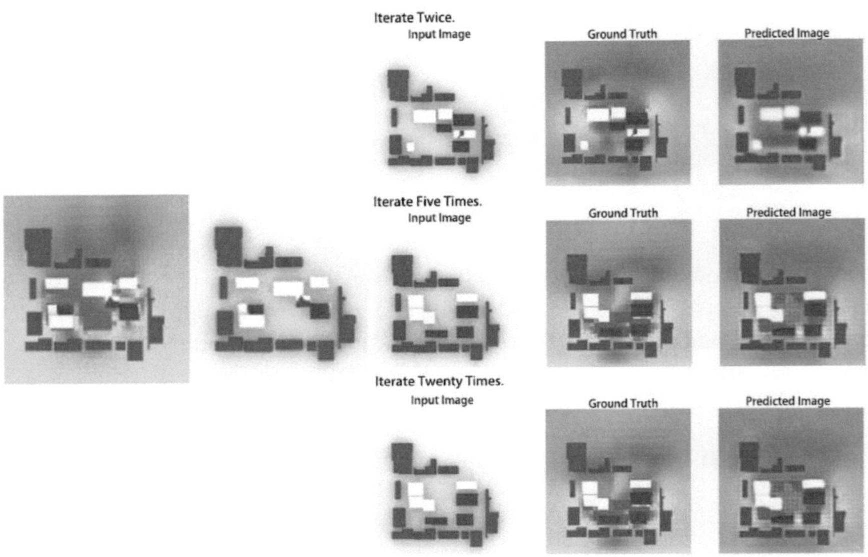

In the wind environment simulation, wind speed and direction simulation results train the Pix2Pix model for wind environment prediction. Through multiple iterations, each evaluates wind flow patterns using the prediction model and compares them with actual simulation results, optimizing the building layout and enhancing ventilation performance.

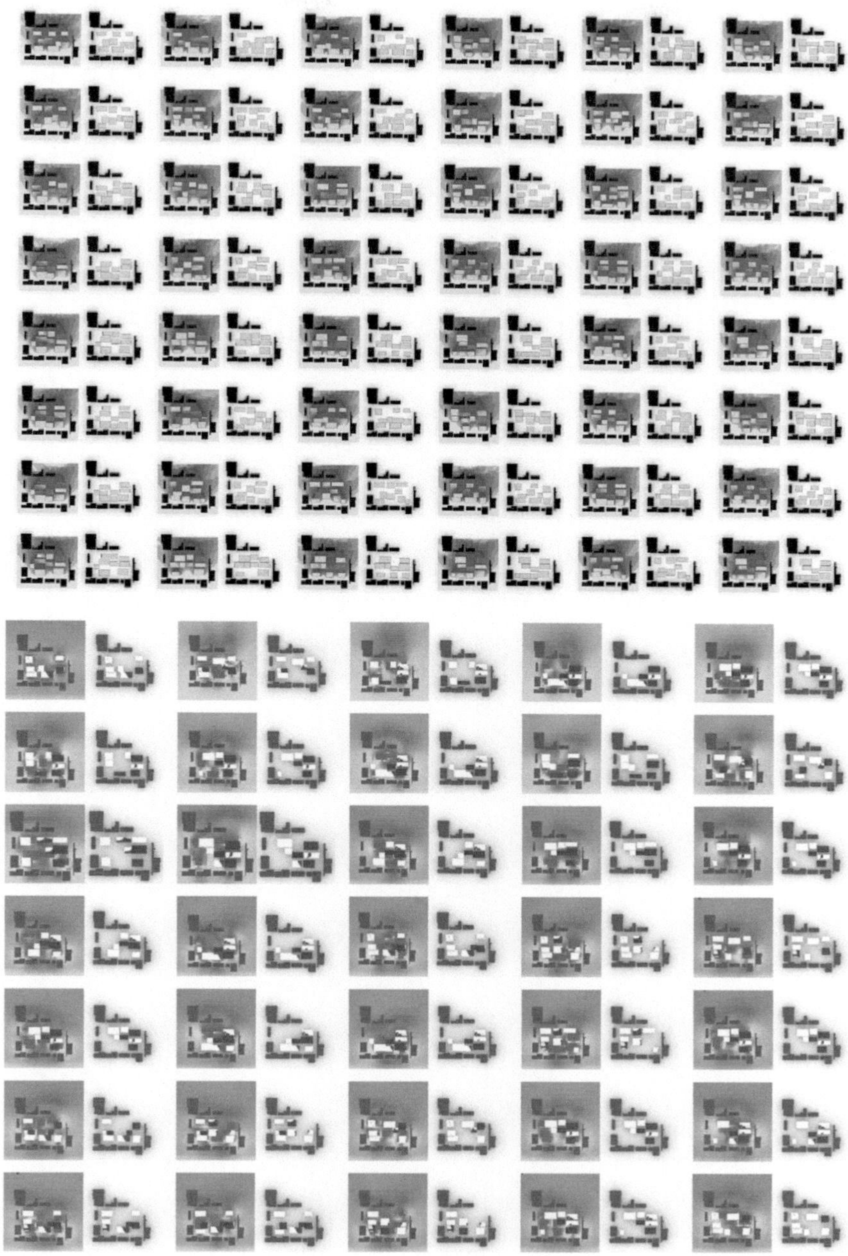

This image shows the training set used for performance prediction.

Chapter 8
Overlay Ecology and the Enjoyment of the New City

Members
He Weili, Wang Chidi, Tang Xu, Ji Yunzhu, Hao Tongping, Peng Huiwang, Wang Hongxin

Background
Exploring urban spatial forms is essential for city characteristics, and urban design is crucial for improving urban environments. Located in Shenzhen's shallow sea area, the project focuses on addressing environmental issues caused by the hot climate. We aim to innovate by using novel foundational logic to generate and optimize architectural forms. Employing geometric stacking methods and integrating simulations into machine learning streamline the process, offering more optimization possibilities.

Method
1. Morphogenetic strategy

The generation logic starts with dividing a plot's boundary into primary blocks, which are then subdivided into six blocks of set areas. These six blocks undergo a secondary division. Buildings are generated within the resulting grid paths, offering diverse architectural forms with options like no generation, one floor, or two floors, allowing for potentially elevated structures.

2. Simulation

The optimization training process begins with randomly generating numerous architectural forms, distinguished by color. Sunlight and wind environment simulations are then conducted on these generated forms. Building parameters and environmental indicators are used as input parameters for machine learning. Next is multi-objective optimization, with four evaluation criteria: plot ratio, solar radiation, UTCI (summer outdoor thermal comfort index), and landscape visibility.

J. Yao, *Intelligent Exploration in Sustainable Design*,
https://doi.org/10.1007/978-981-96-6743-7_8

Due to time constraints, only about 10 iterations were performed, resulting in hundreds of data points. An optimization direction was derived from these iterations, and the best results from different iteration counts were selected. The results show gradual improvements in architectural forms in terms of visibility, daylighting, ventilation, and other conditions.

Significance

The entire project primarily aims to improve urban environments while providing corresponding urban survival guidelines. It strives to achieve rapid assessment and automated generation, offering a possibility for exploring more rational designs in the future within a short timeframe.

8.1 Technology Method

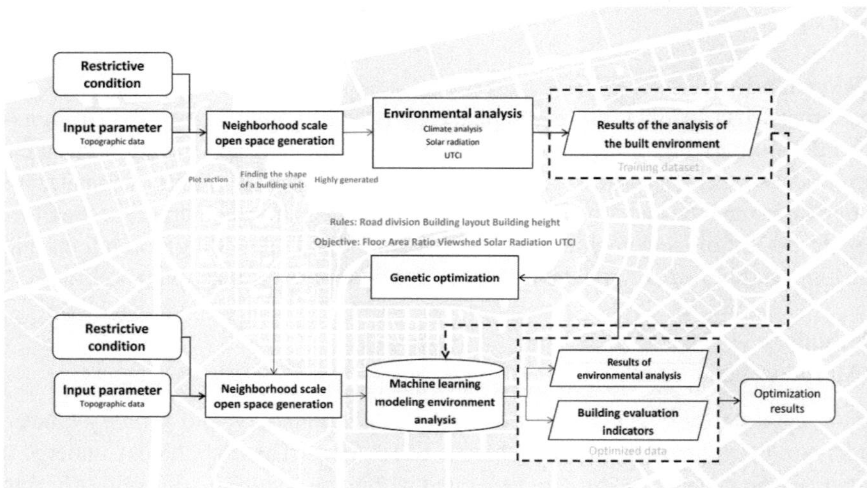

This diagram shows a city design optimization method combining machine learning and genetic algorithms. First, neighborhood-scale open spaces are generated using topographic data and constraints, while environmental analysis (e.g., climate analysis, solar radiation, UTCI indicators) evaluates the built environment, forming the training dataset.

Next, the machine learning model predicts building design performance, eliminating the need for repeated complex simulations. Using these predictions, the genetic algorithm optimizes the building layout to meet criteria like floor area ratio, viewshed, solar radiation, and UTCI indicators. Finally, optimized data assesses the environmental adaptability of the design, enabling efficient and precise urban design optimization.

8.2 Generation Strategy

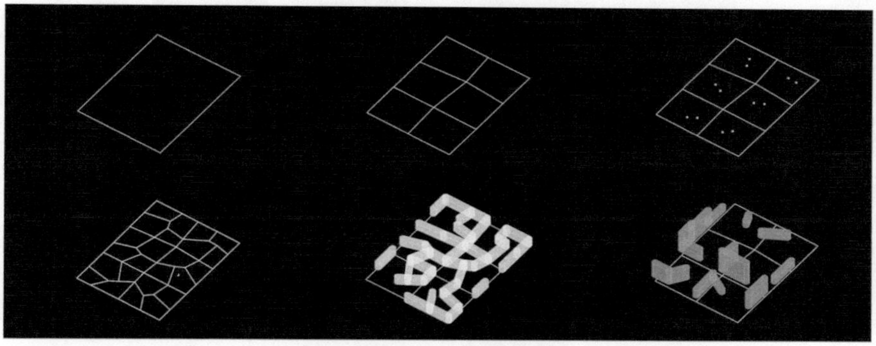

This part demonstrates a building layout generation strategy. It starts with a basic grid layout (shown in the top left), and then the grid is divided into smaller areas. Next, the layout of these areas becomes more complex, gradually forming a strip-shaped building by arranging and combining them in a reasonable way.

8.3 GAN Training

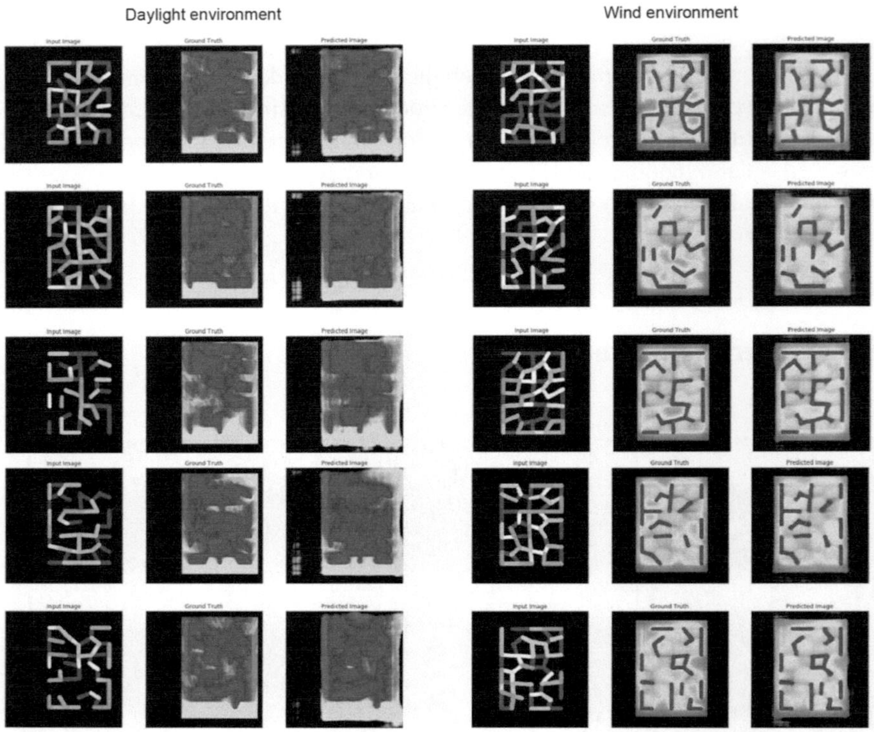

This section explains how a GAN is trained to simulate daylight and wind environments. The image shows building layouts with colored bars for heights. As training progresses, the model learns to predict sunlight and wind flow. Some layouts have raised ground floors to improve light and ventilation. Through multiple cycles, the GAN enhances accuracy, making environmental performance simulation more effective.

8.4 Optimization of Sight Analysis

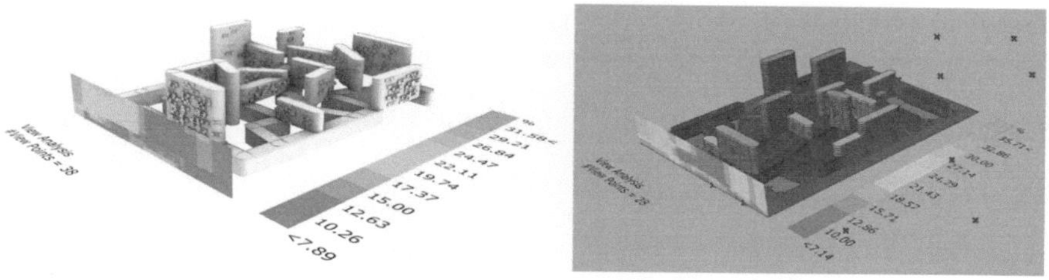

This section demonstrates the optimization process of sight analysis. The images show that different colored areas represent the visibility range of the buildings, with the color variation reflecting the degree of view optimization. The left image shows the sight analysis before optimization, while the right image shows the result after optimization. By adjusting the building layout, the permeability of sightlines and the quality of the views are maximized.

8.5 Optimization of Daylight

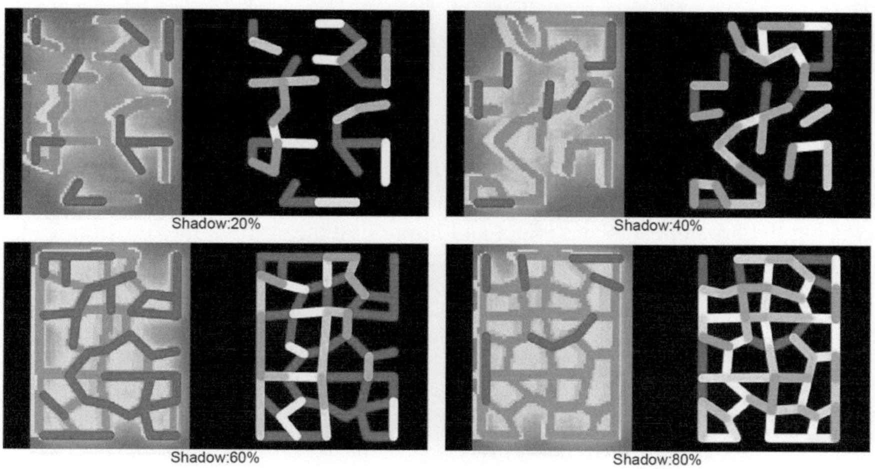

In the daylight optimization section, the images show the optimization of building layouts under different shadow percentages. By adjusting the building arrangement and the extent of shadow coverage, different sunlight conditions can be achieved. Each image compares varying shadow ratios (20%, 40%, 60%, and 80%) to illustrate how to optimize sunlight conditions to enhance environmental comfort.

8.6 Optimization of Heat

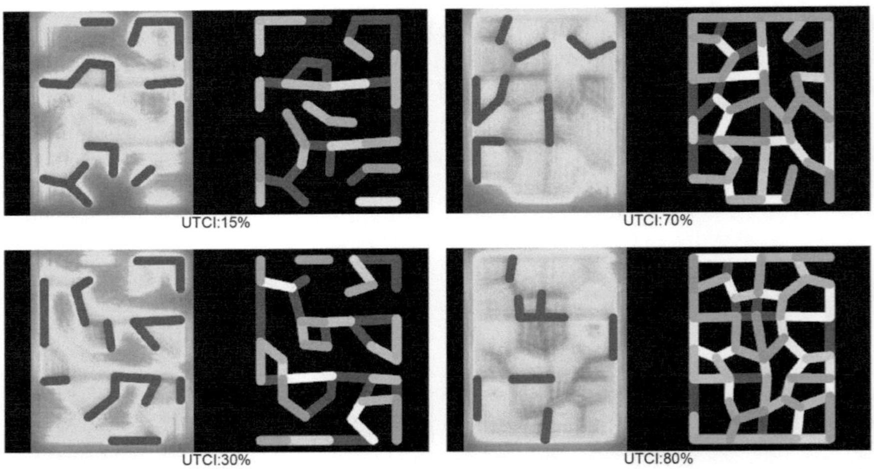

In the heat optimization section, the images display the optimization of building layouts under different thermal conditions (UTCI 15%, 30%, 70%, 80%). By adjusting the layout of buildings and green spaces, the thermal comfort is optimized, reducing the urban heat island effect and improving the surrounding environmental comfort.

8.7 Multiobjective Optimization

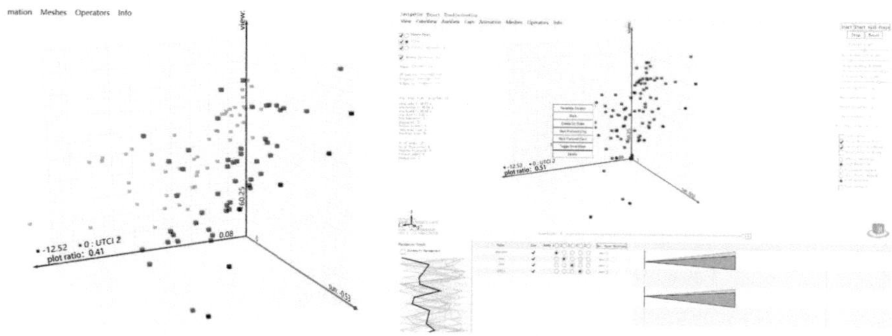

This section demonstrates the multi-objective optimization using the Octopus plugin. The Octopus plugin incorporates genetic algorithms and simulated annealing to help users find the best solutions in complex design spaces. The optimization results are presented using 3D graphics, assisting in balancing and decision-making between multiple objectives.

8.8 Randomized Generation

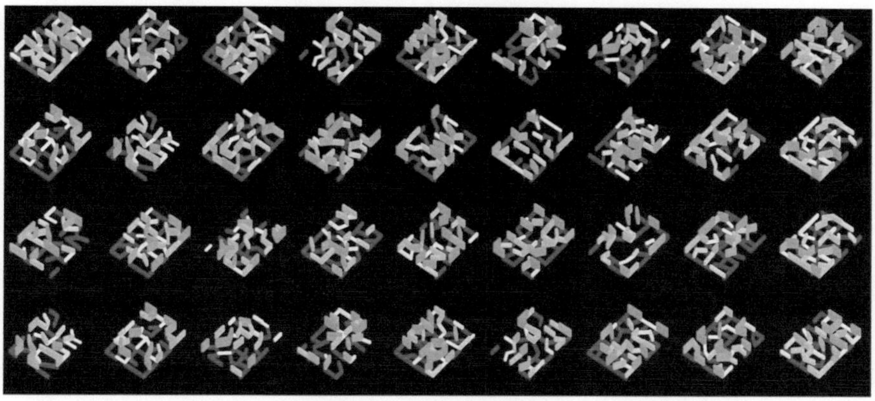

This section demonstrates a randomized generation approach for creating various building layouts. Each layout is represented as a grid, with the green and purple areas indicating different building volumes or heights. By randomizing the generation of these building volumes, a variety of urban design options are produced, which can then be evaluated for performance optimization.

8.9 Results

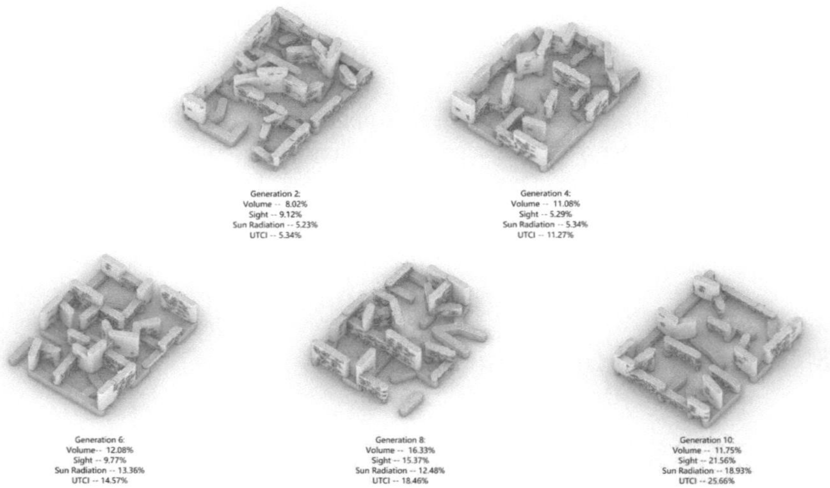

This section displays optimization results across multiple generations. Each generation's building layouts improve in volume, sight, solar radiation, and thermal environment (UTCI). For instance, from generation two to generation ten, building volume and sunlight conditions enhance, and the thermal environment improves progressively. These results show how multiple design criteria are balanced, achieving better environmental and functional performance.

8.10 Final Program

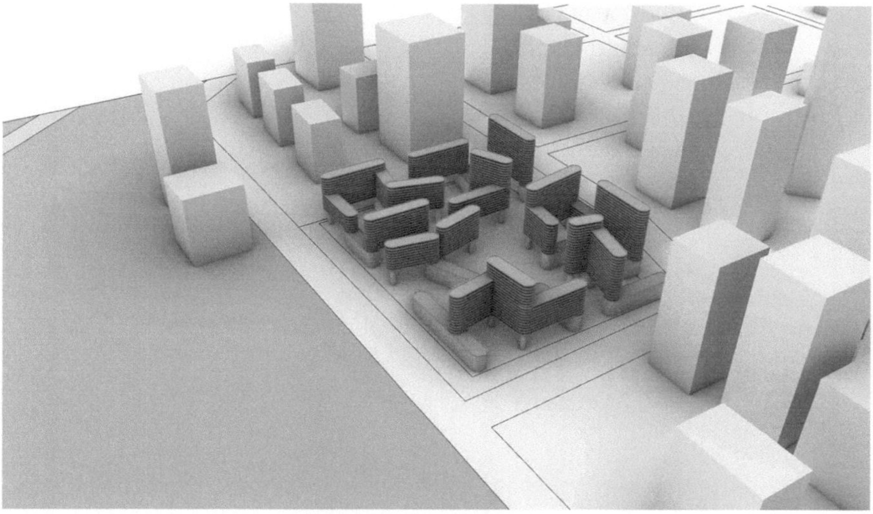

The final design presents a staggered building layout.

Chapter 9
Facing in Light: Biomimetic Facade Generation for Existing Buildings in Different Urban Environments

Members
Zheng Yun, Gan Xinran, Duan Shufan, Zha Xinyu, Huang Tingting

Background
This study aims to tackle lighting issues in urban buildings, like insufficient design standards in older structures or excessive glazing in glass curtain walls. We propose using building facades to address these problems. While dynamic facades can regulate indoor comfort based on environmental changes, they often incur high maintenance costs and energy consumption. Our approach involves analyzing the environment and generating facades optimized for current conditions to maximize indoor lighting.

Method
1. Form Inspiration

We drew inspiration from nature, considering forms such as anemones, resin, and slime molds. Eventually, we opted for the open-mouthed design of sea anemones for the facade. Our site selection included Taipei, Tianjin, and Hangzhou, each with unique environmental characteristics. Taipei, with its dense old city area, demands strong shading in summer. Tianjin's multi-story old office buildings require winter daylighting despite surrounding obstructions. Meanwhile, Hangzhou features buildings with glass curtain walls in a semi-developed area.

2. Daylight Simulation

Based on the site's environment, daylight calculations are used to decide the density of the facade framework, affecting the size of its openings and protrusion degree. Key

J. Yao, *Intelligent Exploration in Sustainable Design*,
https://doi.org/10.1007/978-981-96-6743-7_9

optimization parameters include the size of each facade unit's opening and surface curvature. By adjusting these parameters, the facade's morphology is altered to optimize the proportion of adequately shaded areas indoors. Uneven grids generated from sunlight data are scaled vertically to achieve the optimal facade morphology.

Significance

This exploration in generating building facades considers climate, environment, and architectural conditions. It emphasizes facade design, strongly linked to building aesthetics and energy consumption. Performance-based generative exploration may become a model for future facade design research.

9.1 Technology Method

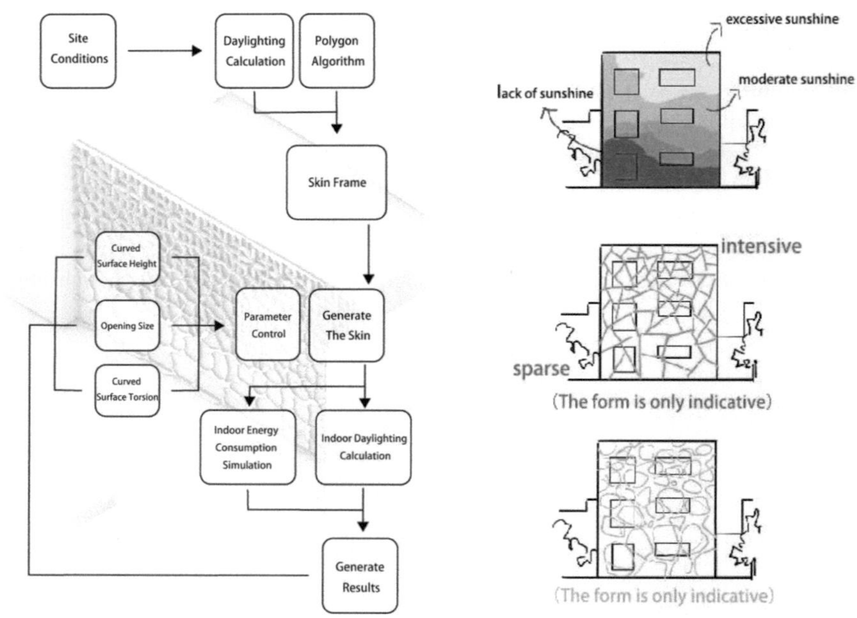

This method optimizes the building facade (skin) based on site conditions and daylighting calculations. A polygon algorithm generates a skin frame with parameters like curved surface height, opening size, and surface torsion. These parameters simulate indoor energy consumption and daylighting conditions, guiding the outer shell design. The result is a facade adapting to varying sunlight conditions—balancing excessive, moderate, and lacking sunlight. The skin design is fine-tuned through these parameters to ensure optimal indoor lighting and energy efficiency.

9.2 Inspiration

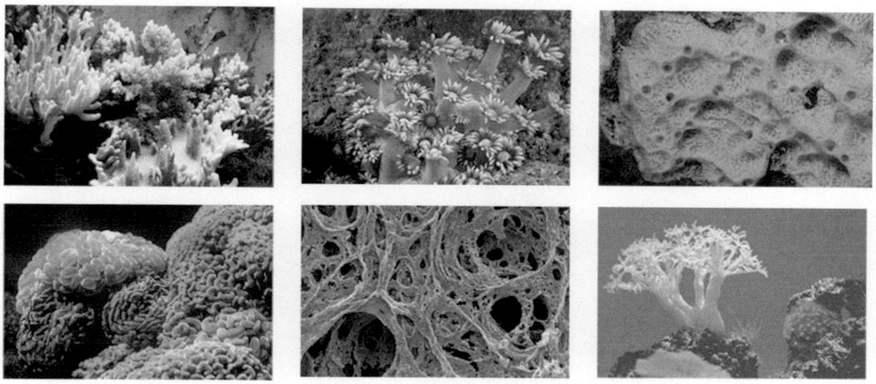

These images of microscopic marine life and coral structures inspire the building's facade design. Their layered, complex, and porous forms enhance shading, daylighting, and natural ventilation, optimizing light and airflow to improve energy efficiency and comfort.

9.3 Facade Exploration

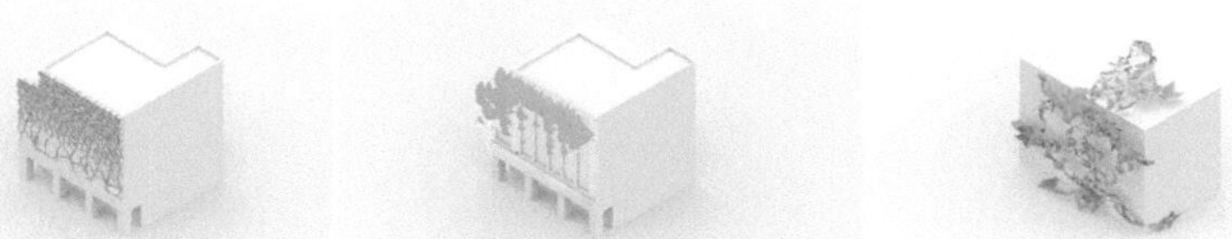

This section demonstrates the exploration of different facade forms. Various facade designs with potential for shading and ventilation were created. By altering the structure's density, layering, and porosity, different forms were explored to optimize the facade design, laying the foundation for subsequent performance analysis and optimization.

9.4 Daylight Simulation

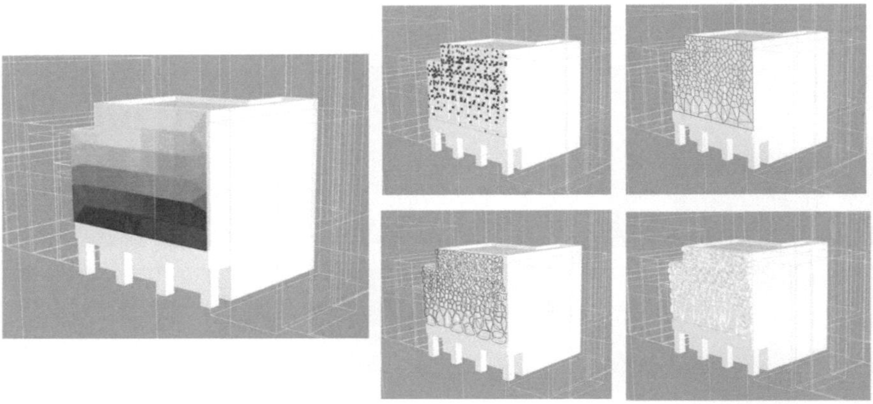

This section shows daylight simulation results for different facade forms. The simulations visualize how sunlight interacts with the facade, with varying levels of sunlight exposure across the building surface. The forms with different structural densities and porosities are assessed to understand how these variations influence the distribution of natural light inside the building, guiding the optimization of facade design for improved daylight access.

9.5 Facade Generation and Optimization

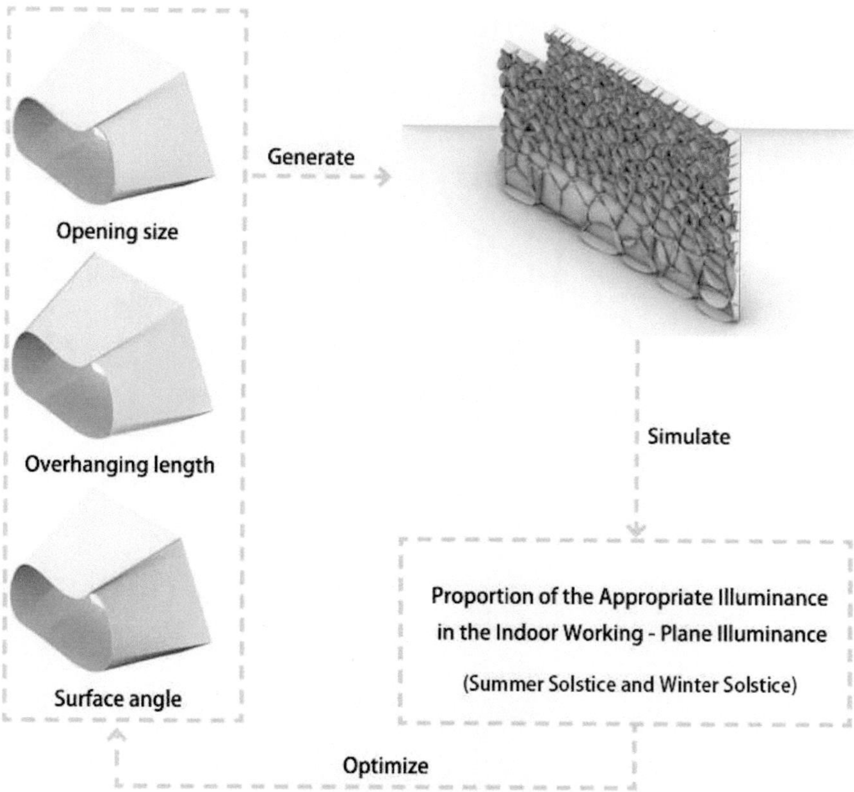

This part demonstrates the process of generating and optimizing the building facade. First, the form of the facade is generated by adjusting the opening size, cantilever length, and surface angles. Then, simulations are conducted to assess the indoor daylight ratio under different designs, analyzing the suitable lighting conditions during the summer solstice and winter solstice. Finally, based on the simulation results, optimization is carried out to ensure indoor lighting comfort and energy efficiency.

9.6 Results of Simulation

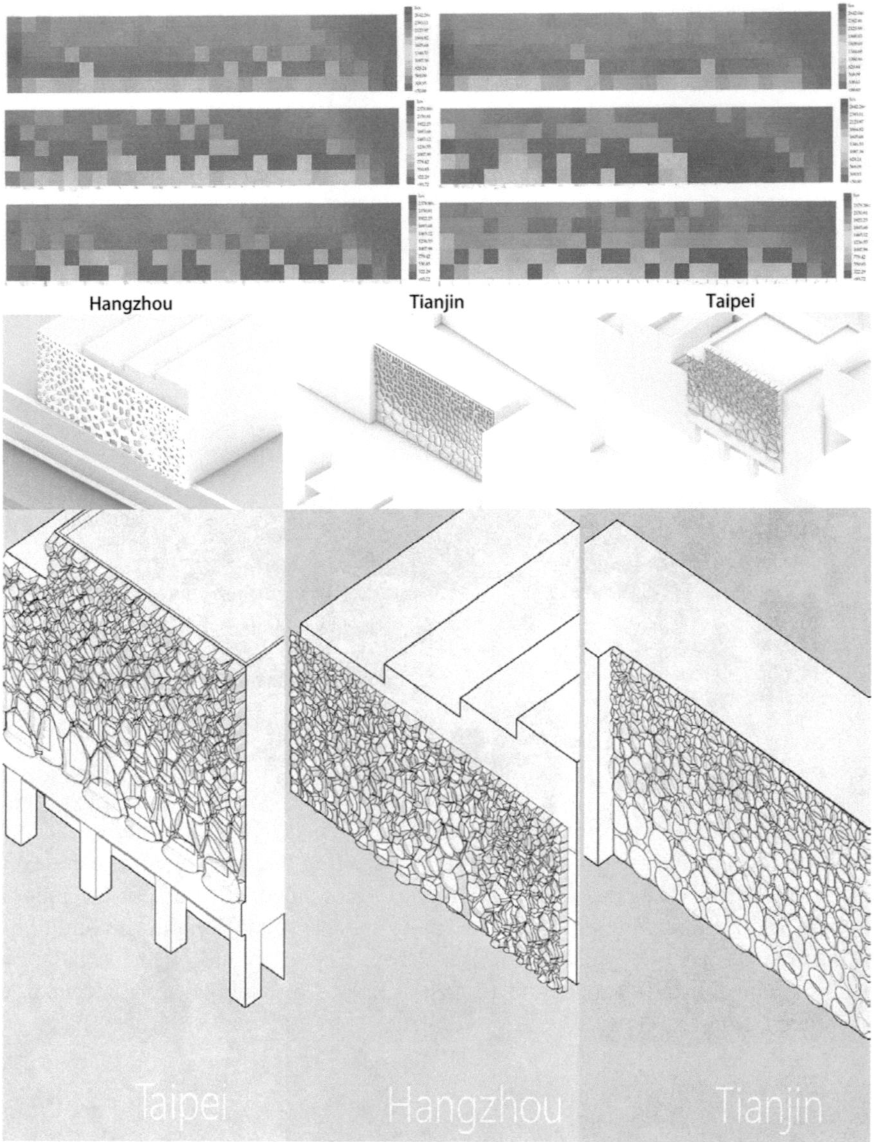

This part shows the simulation results of building facades in three different cities: Hangzhou, Tianjin, and Taipei. By comparing the simulation data from different cities, it is possible to observe how different facade designs perform under various climate conditions.

Part II
DigitalFUTURES 2022: AI for Carbon-Neutral City

Mentors
Yao Jiawei, Meng Jing, Fu Bin, Huang Chenyu, Zhang Gengjia, Yin Minggang, Anna, Li Jingyi

Assistants
Wang Jinyu, Wang Xiaoxin

Members
Guo Fang, Wang Changsi, Li Runtian, Gu Youlong, Wu Qilin, Chen Wenzhen, He Yanze, Lei Zhen, Liu Yiming, Liu Yuqiu, Zhuang Junling, Chen Yutong, Jiang Qingrui, Li Danrui, Luo Zhaoyang, Zhang Hanqi, Cui Qinyu, Zhang Chi, Feng Jiajia, Bi Xuejiao, Wang Lei, Lu Yijun, Yu Mengqing, Chen Qiyue, Wu Zihao, Yu Bingqing, Wen Wen, Zhang Yecheng, Qi Ziying, Xiong Hongyi, Wang Zixuan, Xu Wenya

Topic
In response to the current climate change, China has set development goals for peaking carbon dioxide emissions by 2030 and achieving carbon neutrality by 2060. Currently, cities consume two-thirds of the global energy and generate over 70% of carbon emissions. As designers, how can we integrate design and technology to create carbon-neutral cities? As researchers, how can we predict urban carbon emissions and develop roadmaps for achieving carbon neutrality? Artificial intelligence may help us find answers. This workshop will be divided into two directions: future design and intelligent deduction, exploring how artificial intelligence can assist in achieving carbon-neutral cities from both design and research perspectives.

Content
In the Future Design track, three groups of five participants each will explore intelligent design methods for achieving carbon-neutral cities through a performance-driven design workflow.

In the Intelligent Deduction track, three groups of five participants each will study carbon emission prediction research for cities and buildings using geographic information systems, multi-source data, and urban energy modeling, and provide recommendations for urban development directions.

Trainee Information in 2022

180%	Total applicants
43%	Have experience in artificial intelligence applications
56%	Have published academic papers
57%	Have research experience related to carbon neutrality
60%	Have experience related to performance-based design

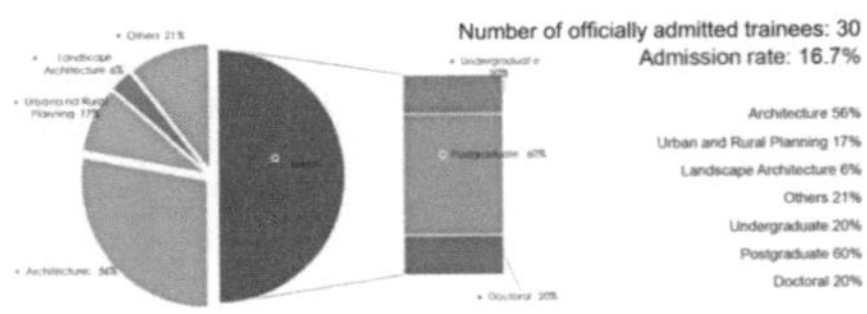

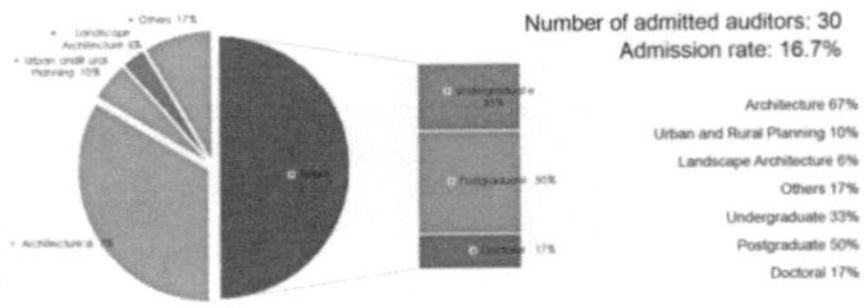

Paper Publication

·An N, Huang C, Shen Y, et al. Challenges of carbon emission reduction by the workshop education pattern[J]. Heliyon, 2023, 9(3) e13404.

·Lu Y, Chen Q, Yu M, et al. Exploring spatial and environmental heterogeneity affecting energy consumption in commercial buildings using machine learning[J]. Sustainable Cities and Society, 2023, 95: 104586.

·Jiang Q, Huang C, Wu Z, et al. Predicting building energy consumption in urban neighborhoods using machine learning algorithms[J]. Frontiers of Urban and Rural Planning, 2024, 2(1): 6.

Chapter 10
Optimal Design of GAN Generative Master Plan Based on Plant Carbon Sink Performance and Building Energy Consumption Calculation

Members
Guo Fang, Wang Changsi, Li Runtian, Gu Youlong, Wu Qilin, Chen Wenzhen

Background
Existing research on carbon emission calculations primarily focuses on buildings and roads, neglecting vegetation in carbon sequestration within cities. While traditional studies acknowledge the role of vegetation in carbon emissions, energy consumption optimization tends to prioritize the spatial arrangement of vegetation between roads and buildings. Therefore, this project aims to coordinate optimization among the "building-road-plant" elements, constructing a set of overall designs with better carbon emission predictions and sunlight analysis.

Method
1. Building, road, and vegetation layout

Collect high-definition satellite images of Wuhan and manually annotate three city elements (80 images). Supplement with mask-RCNN semantic segmentation (70 images). Apply data augmentation to expand the dataset to 600 images. Train a PIX2PIX model with this dataset. Use the trained model to predict vegetation layout in the city, generating greenery on plain maps while preserving Wuhan's urban texture.

2. Residential layout optimization

In Grasshopper using Python, develop a building density optimization program. Set constraints like "plot ratio," "density," "height limit," and "building lines." Inputs

J. Yao, *Intelligent Exploration in Sustainable Design*,
https://doi.org/10.1007/978-981-96-6743-7_10

include "number of buildings," "number of floors," and "building area." Incorporate the PIX2PIX greenery prediction model to generate layouts for "road-building-plant" arrangement, guiding carbon sink and emission optimization.

3. Multi-objective optimization in GH.

Deploy the trained model on GH platform to generate 3D block models. Use carbon emission calculation formula for "road-building-plant" triad. Conduct multi-objective optimization for carbon sink capacity, road block energy consumption, daylighting performance, with Isight genetic algorithm. Variables include "building rotation," "central plaza presence," "cross-distribution method," and "position adjustment." Goal is to find layout with superior environmental performance while retaining Wuhan's urban texture.

Significance
From a green carbon reduction perspective, while preserving the original urban block texture, here are some strategic suggestions for green layout in the early stages of architectural design.

10.1 Technology Method

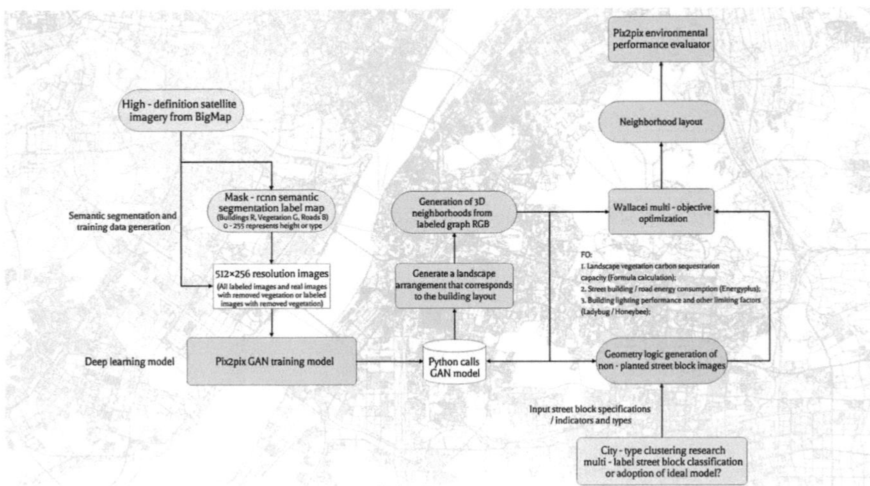

This diagram outlines the process of creating building and green space configurations using Generative Adversarial Networks (GANs), while optimizing building carbon emissions and green space carbon sequestration.

The process starts with high-resolution satellite imagery and Mask-RCNN for semantic segmentation, extracting data on buildings, green spaces, roads, and other land use types. These segmented images form training datasets, including building height, type, green spaces, and roads.

Next, the Pix2Pix GAN model is trained to produce 3D neighborhood layouts that resemble real urban environments. Using these layouts, Python calls the GAN model to generate street block configurations.

Subsequently, the Wallacel multi-objective optimization tool optimizes building layouts, considering factors like building carbon emissions, green space carbon sequestration, street and building energy consumption, and more. This step also involves city-type clustering research for street block classification and selecting an ideal model based on these classifications.

Finally, the generated street blocks are analyzed for their performance in carbon emissions, green space carbon sequestration, and environmental adaptability, offering optimized urban planning solutions.

10.2 Machine Learning Algorithms

mask-RNN

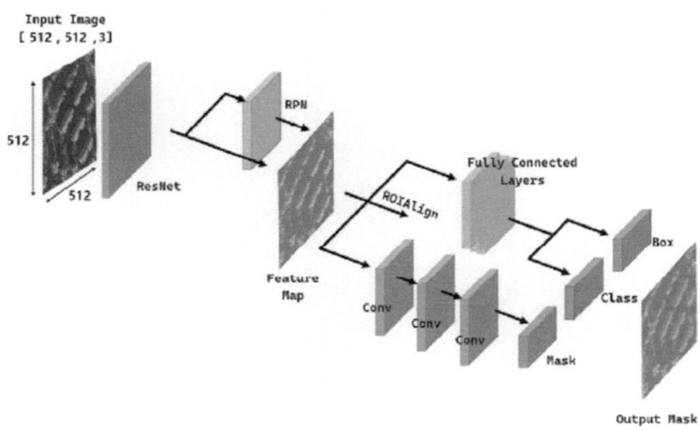

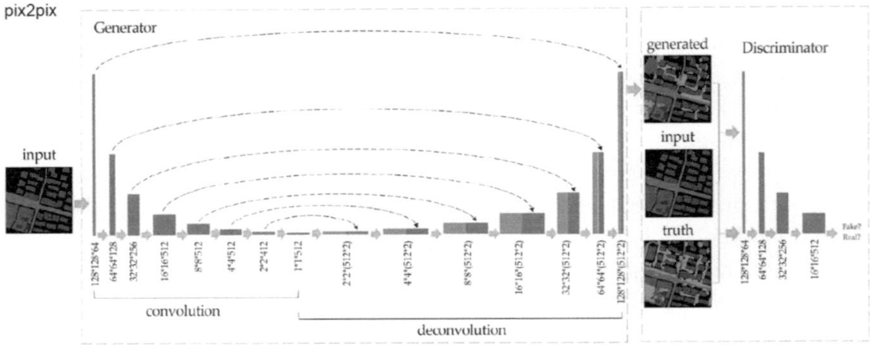

This section presents two machine learning algorithms for generating building and green space configurations: Mask-RNN and Pix2Pix.

First, Mask-RNN performs image semantic segmentation. The input image is processed via a ResNet network, extracting features through feature maps. Convolution and fully connected layers then produce target bounding boxes and segmentation masks. This algorithm identifies and segments areas like buildings, green spaces, and roads, providing data for subsequent generation and optimization.

Next, Pix2Pix is introduced, where the input includes buildings and streets, and the output is green space distribution. Together, these algorithms optimize building and green space layouts, enabling precise and adaptable urban design through deep learning.

10.3 Selected Block

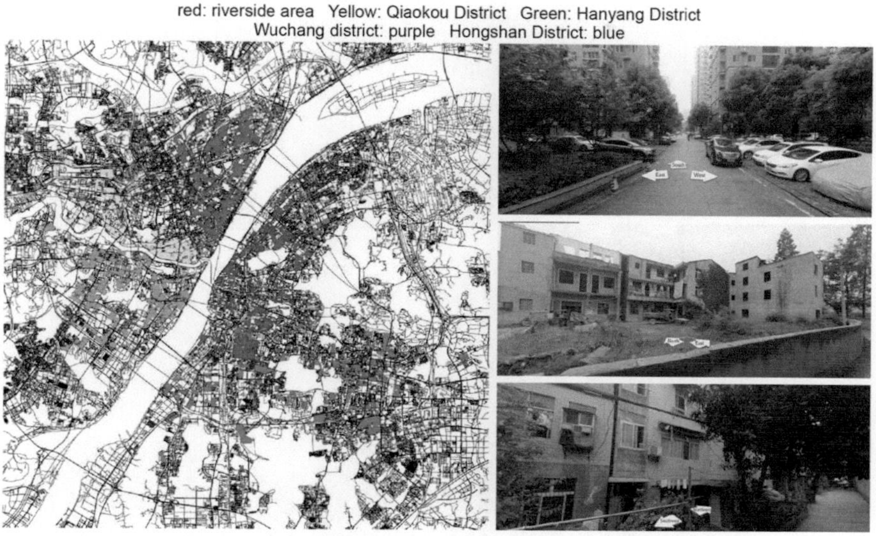

This section shows the selected blocks for data collection, including real-world photographs of the streets and buildings, providing actual geographical and environmental data for subsequent analysis and design.

10.4 LabelMe Segmentation

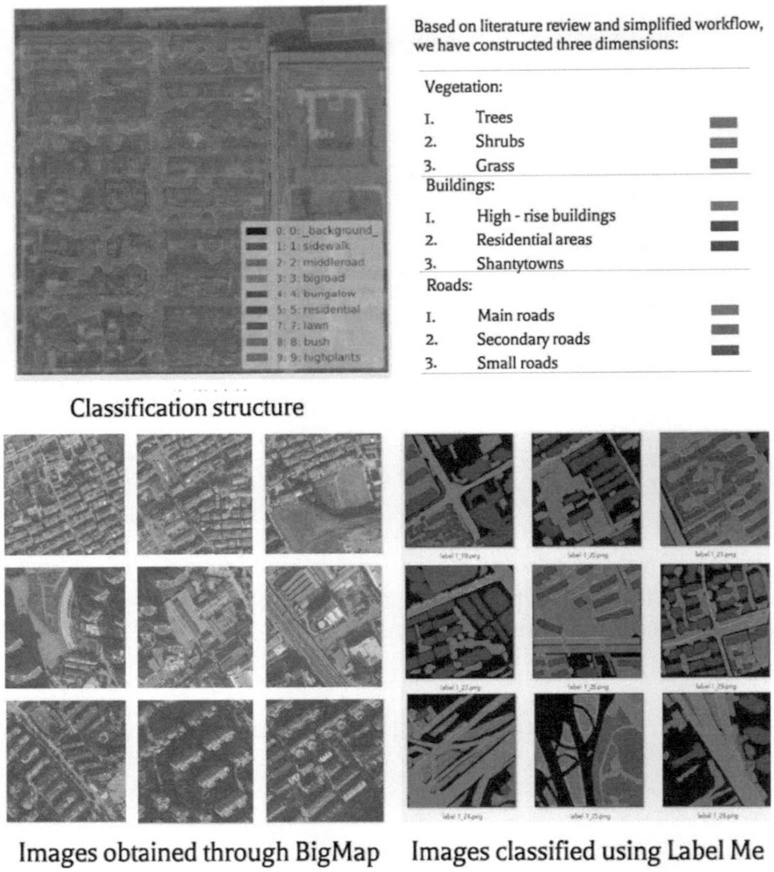

This section outlines the data annotation process via image segmentation. Satellite images are first acquired through BigMap for initial classification, followed by detailed annotation using LabelMe to categorize elements like roads, buildings, and vegetation. The classification structure comprises three main categories: vegetation (trees, shrubs, grass), buildings (high-rise, residential, shantytowns), and roads (main roads, secondary roads, small roads).

10.5 Pix2Pix Model Deployment

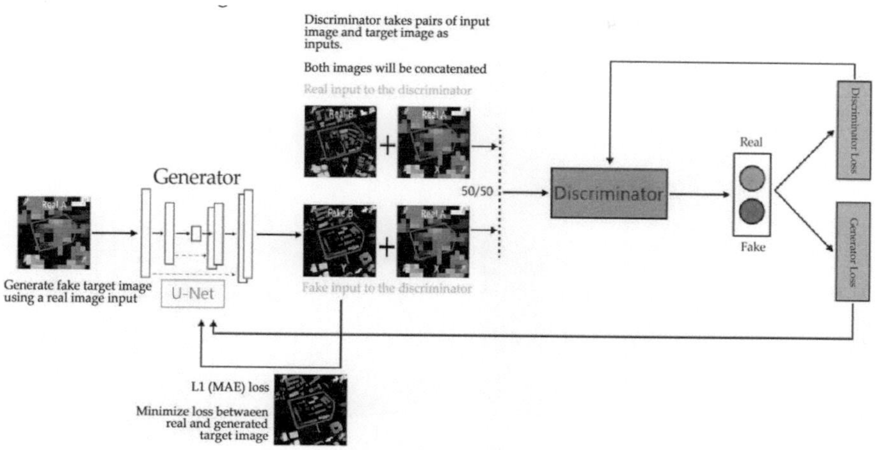

This section explains the Pix2Pix model's operation. The generator produces synthetic images, while the discriminator evaluates their authenticity. Through iterative training, the generator refines its output to closely mimic real images, enabling high-quality image generation.

10.6 Data Amplification

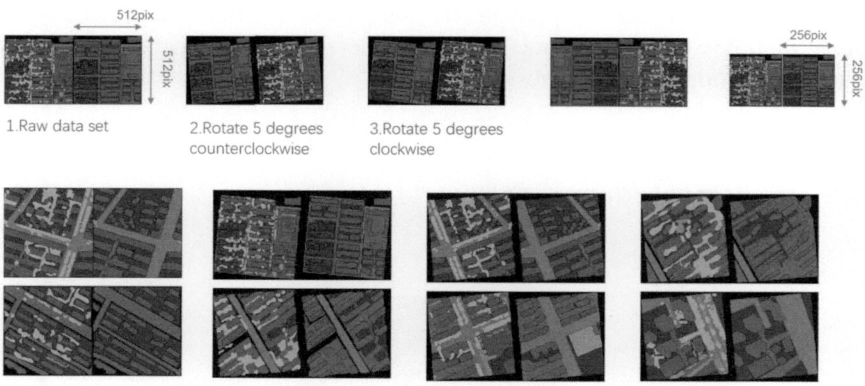

This section illustrates the data augmentation process. Rotation operations, such as rotating images by 5 degrees clockwise and counterclockwise, effectively increase dataset diversity. Additionally, reducing image resolution from 512 × 512 to 256 × 256 pixels enhances data variety and strengthens the training model's robustness.

10.7 Model Training

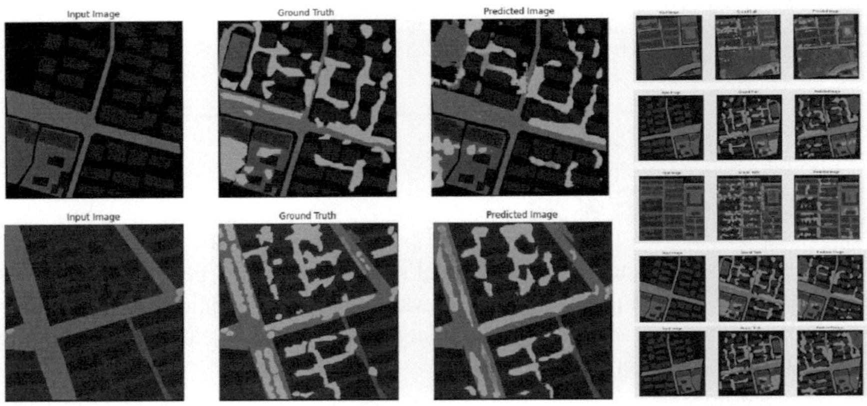

This section outlines the model training process, displaying input images, ground truth, and predicted outputs. Input images represent raw data, ground truth denotes actual classified data, and predicted images are generated by the trained model. The process involves comparing predictions with ground truth to enhance the model's accuracy in classifying urban elements such as roads, buildings, and vegetation. Multiple examples illustrate the model's iterative improvement in accurately segmenting various urban features.

10.8 Settlement Planning

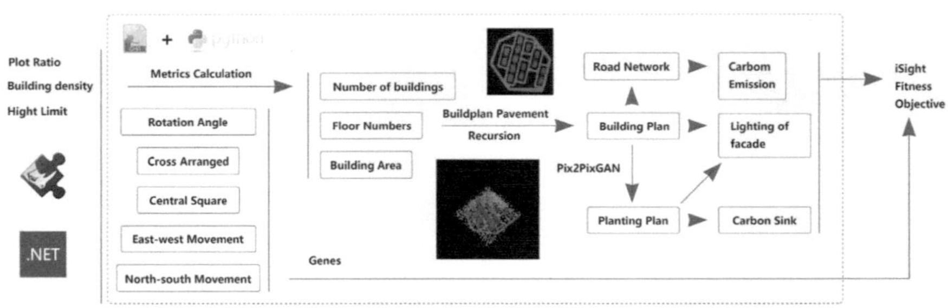

Basic metrics such as building density, height restrictions, and block arrangements are calculated to generate an initial building layout. Recursion is then used to create the pavement and road network. The Pix2Pix model generates a landscaping plan that aligns with the building layout. Design results are evaluated against objectives like carbon emissions, facade lighting effects, and green space carbon sequestration. Finally, iSight software optimizes the design to meet environmental performance standards, enhancing overall urban planning.

10.9 Optimization Results

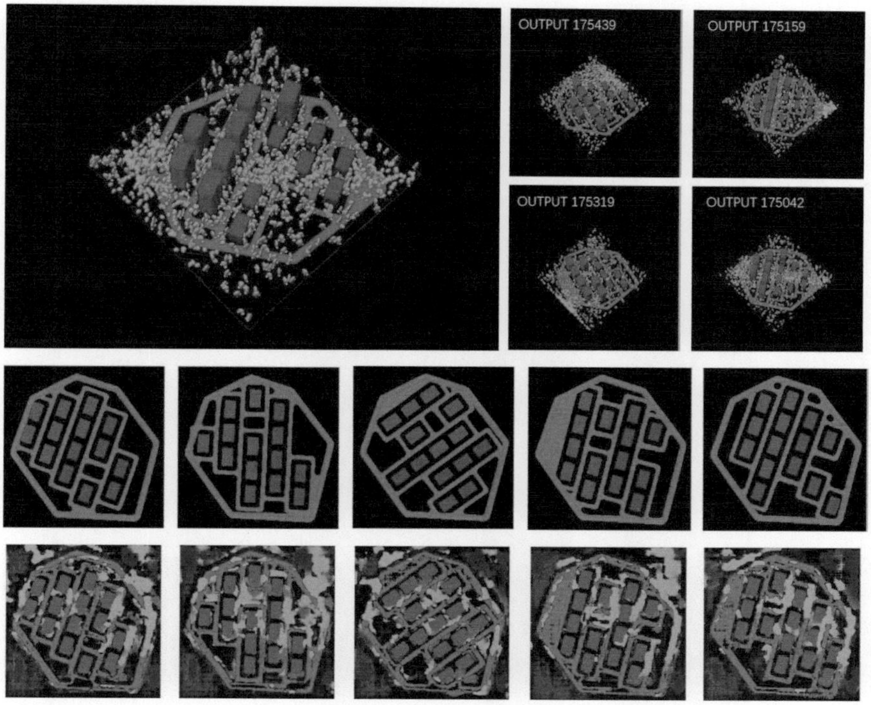

This section outlines how to parameterize and adjust road and building layouts. The designer modifies these layouts based on set parameters before inputting them into the Pix2Pix model to create the corresponding landscape. This process iterates multiple times to optimize the spatial arrangement of buildings, green spaces, and roads for better environmental performance.

Chapter 11
Exploring the "Intelligence" Potential of Rooftop Greenery in Post-pandemic Era

Members
He Yanze, Lei Zhen, Liu Yiming, Liu Yuqiu, Zhuang Junling

Background
During a citywide lockdown in Shanghai, transportation and logistics disruptions hindered food delivery, creating an urban island scenario. This prompted thoughts on ensuring residents' vegetable food security and self-sufficiency. Simultaneously, government-led carbon neutrality policies are pushing for energy efficiency in construction. Rooftop spaces, with solar roof systems for carbon sequestration and rooftop farms for autonomous agriculture, offer a new solution for urban development.

Method
1. Data Collection and Simulation

We gathered data from Shanghai communities affected by the 2022 pandemic, selecting the top five for rooftop solar radiation simulation. Results showed minimal variation in rooftop heights, with higher radiation internally and lower at the edges.

2. Roof Layout Optimization

The combination of photovoltaics and farming brings complementary benefits, like plant transpiration cooling and boosted solar panel efficiency. This study explores

J. Yao, *Intelligent Exploration in Sustainable Design*,
https://doi.org/10.1007/978-981-96-6743-7_11

three rooftop layouts: full coverage, 10-meter modules, and 4-meter modules, to maximize solar energy utilization and enhance temperature exchange efficiency. Economic effects and carbon emissions are key considerations. For photovoltaic power generation, this includes costs, electricity savings, and carbon emissions reduction. Rooftop farming economic effects involve costs, vegetable sales, and reduced building energy consumption, with carbon emissions covering reduced cooling energy use and transportation emissions. The study references existing literature for economic benefits and carbon reduction methods, using the Wallace X multi-objective genetic algorithm on the GH platform to optimize different grid systems for 86 Pareto optimal solutions, balancing carbon reduction and economic effects objectives.

3. Result and Discussion

Different solutions reflected trade-offs between objectives, forming an economic-ecological balance curve. Rooftop function distribution correlated strongly with solar radiation. High-radiation rooftops accommodated more panels, leading to greater carbon reduction and lesser economic benefits. Grid size significantly influenced results, with finer divisions providing better economic and carbon reduction outcomes.

Significance

Propose a low-carbon rooftop system for community self-sufficiency and positive local impacts, balancing economic and ecological benefits through integrated rooftop agriculture and energy. Provide guidance for its design and promotion.

11.1 Technology Method

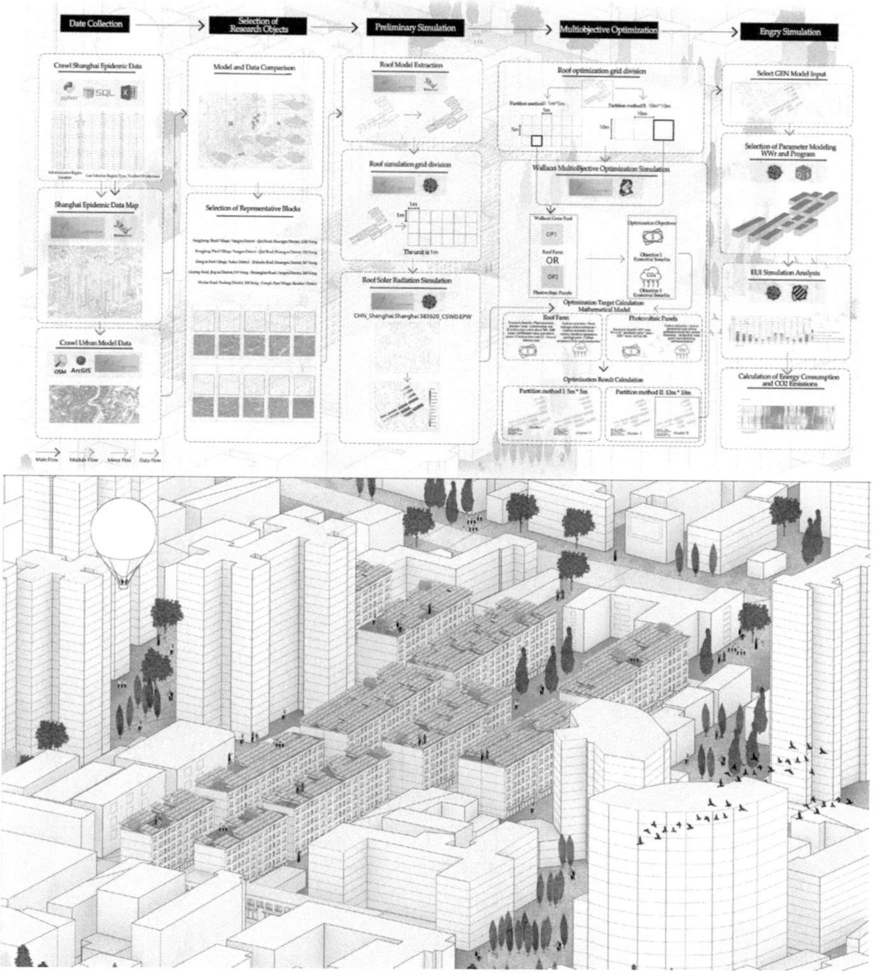

This section outlines a project focused on optimizing rooftop planting and photovoltaics. It begins with data collection and model selection, followed by preliminary simulations including roof extraction and solar radiation assessment. Multi-objective optimization methods are then used to enhance the rooftop layout, while appropriate models are chosen for energy consumption and environmental impact calculations. Through iterative optimization, the best layout for rooftop planting and photovoltaics is achieved, and energy simulations validate its effectiveness.

11.2 Research Background

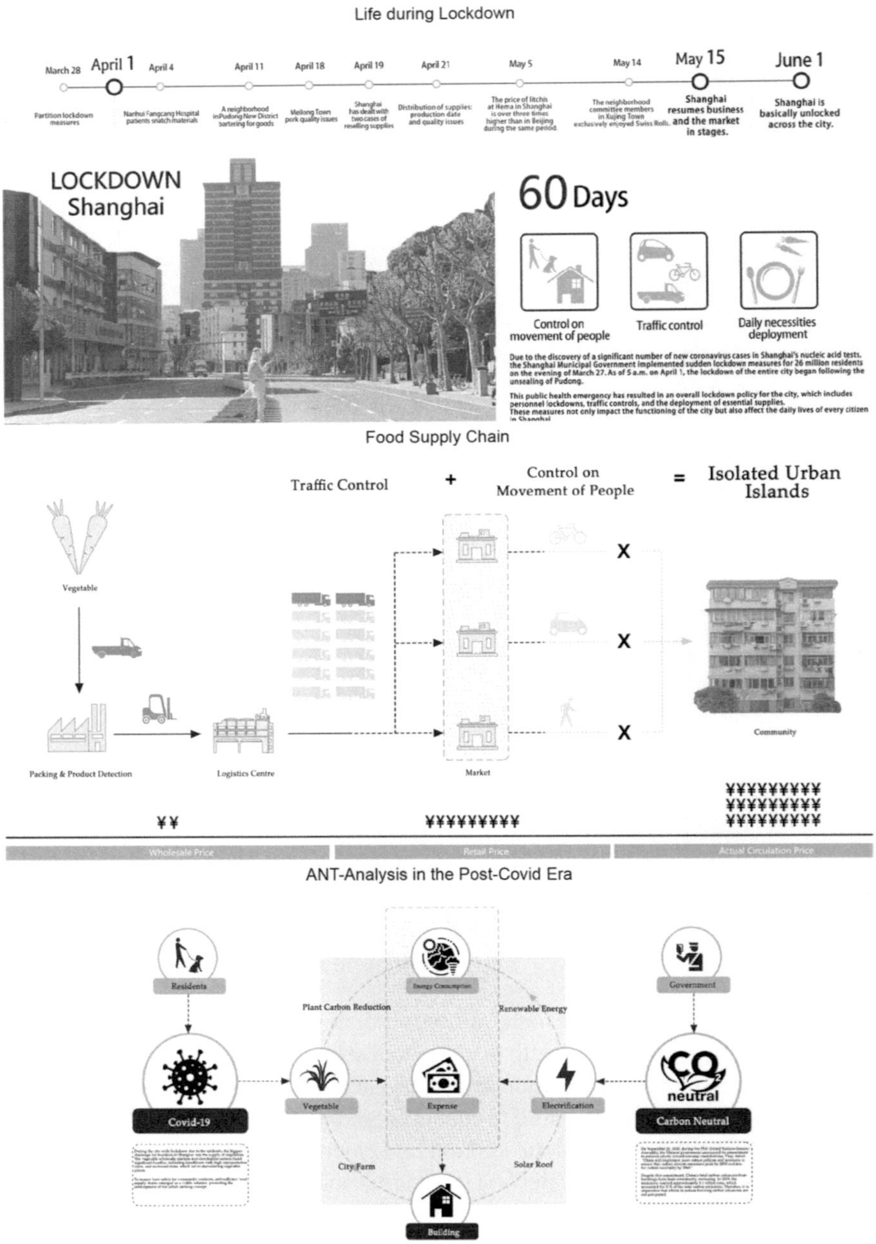

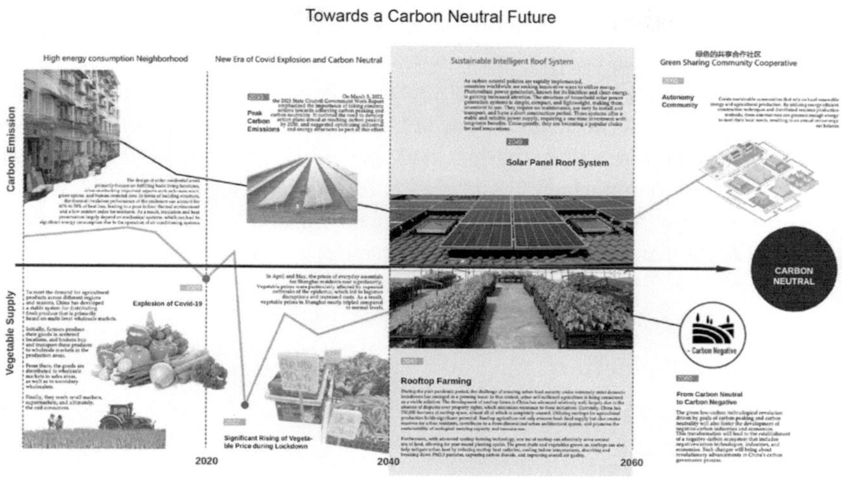

This image depicts the shift from high-energy neighborhoods to a carbon-neutral future, showcasing sustainable urban development with solutions like rooftop solar and farming.

11.3 Evaluation Model

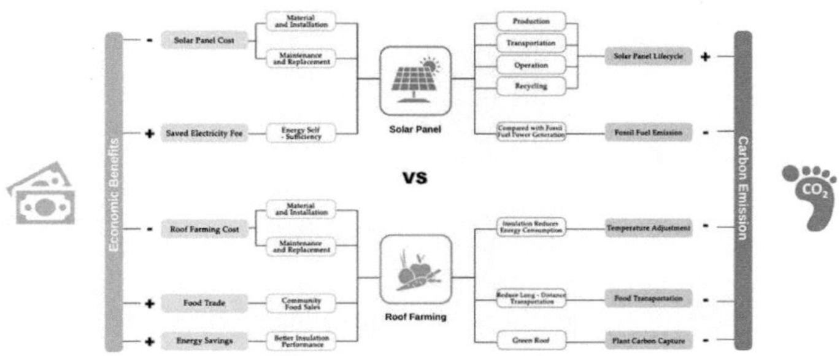

This diagram outlines the evaluation model for rooftop photovoltaics and farming regarding economic and carbon emission impacts. For photovoltaics, costs include installation and maintenance, while savings arise from reduced electricity bills and self-sufficient energy. Carbon emissions are assessed against fossil fuel generation, considering production, transportation, and recycling. For rooftop farming, costs involve setup and maintenance, with benefits from food sales, community trading, and energy savings from insulation and reduced transport. It also aids in carbon sequestration, temperature regulation, and enhances green roof efficiency.

11.4 Sample Selection

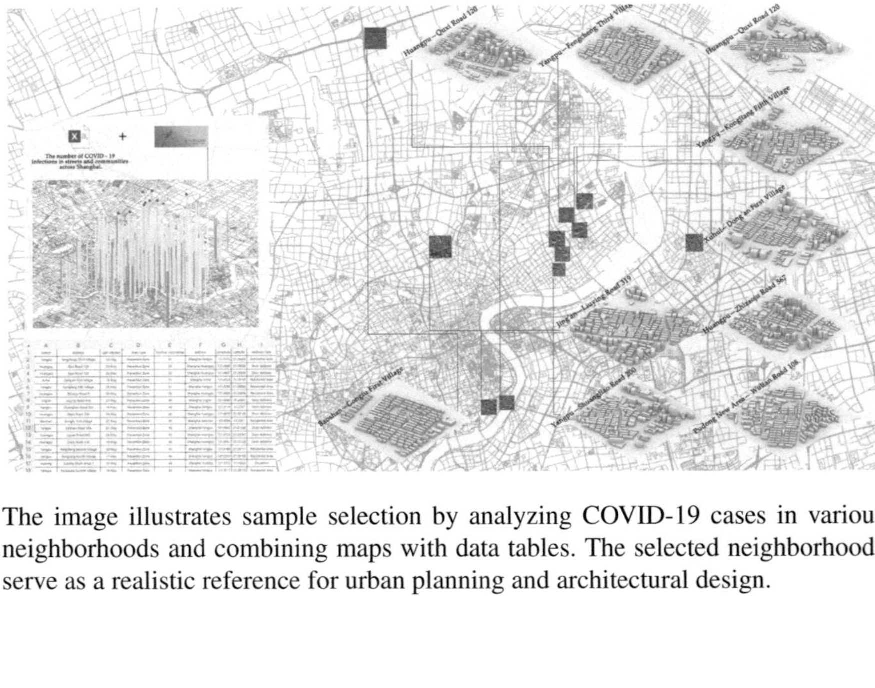

The image illustrates sample selection by analyzing COVID-19 cases in various neighborhoods and combining maps with data tables. The selected neighborhoods serve as a realistic reference for urban planning and architectural design.

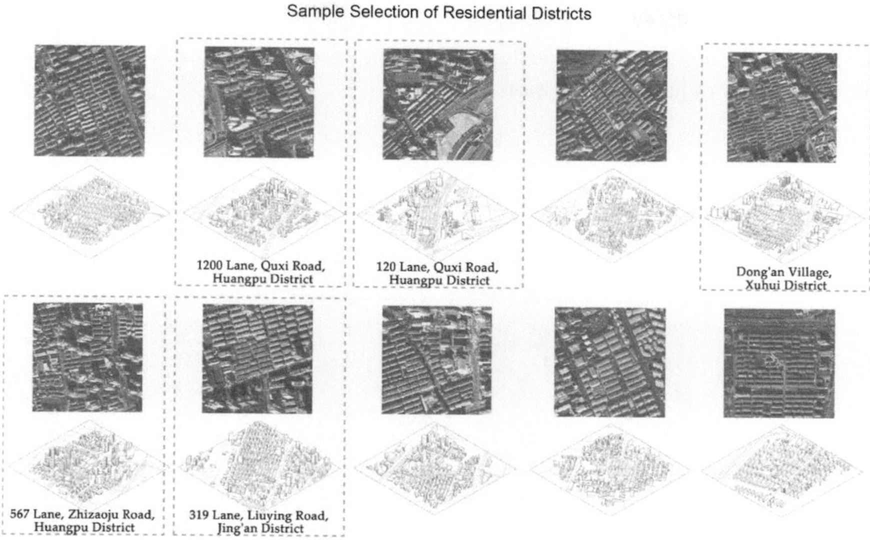

This image shows selected residential districts in Shanghai, featuring aerial views and 3D models of areas like 1200 Lane, Quxi Road in Huangpu District, and Dong'an Village in Xuhui District. These districts were chosen for analysis due to their urban characteristics.

11.5 Analysis of Solar Radiation

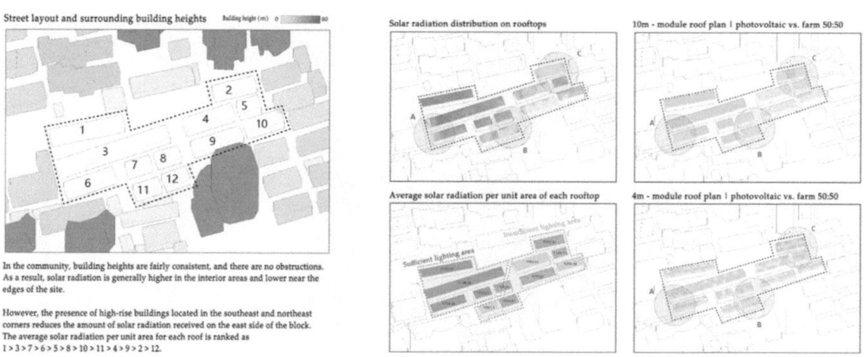

This image analyzes solar radiation on rooftops in a community, showing street layouts and uniform building heights. It indicates higher solar radiation in interior areas and lower near the edges. The map displays the average solar radiation per rooftop area, highlighting sufficient and insufficient lighting. Additionally, it compares two rooftop module designs (50:50 photovoltaic vs. farm ratio) and their effects on solar radiation absorption.

11.6 Roof Meshing

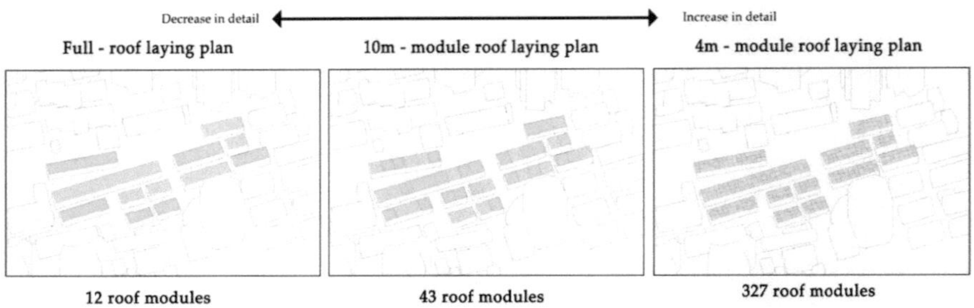

This image illustrates roof meshing at three detail levels. On the left, the full roof layout has 12 modules with less detail. The middle shows a 10-meter module layout with 43 modules, offering more detail. On the right is the 4-meter module layout with 327 modules, providing the most precise design. Increasing the number of modules results in a more detailed roof layout.

11.7 Multi-objective Optimization Settings

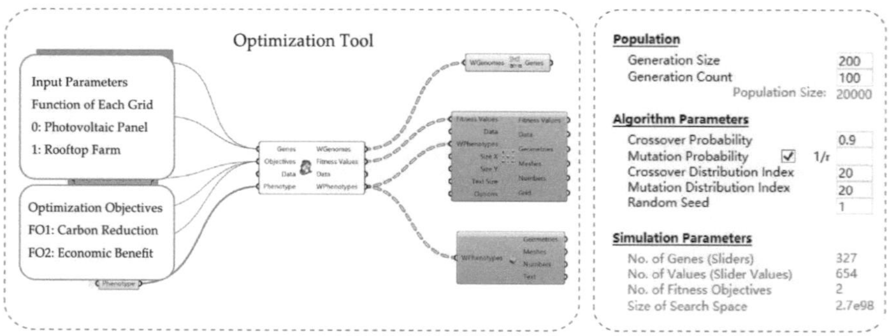

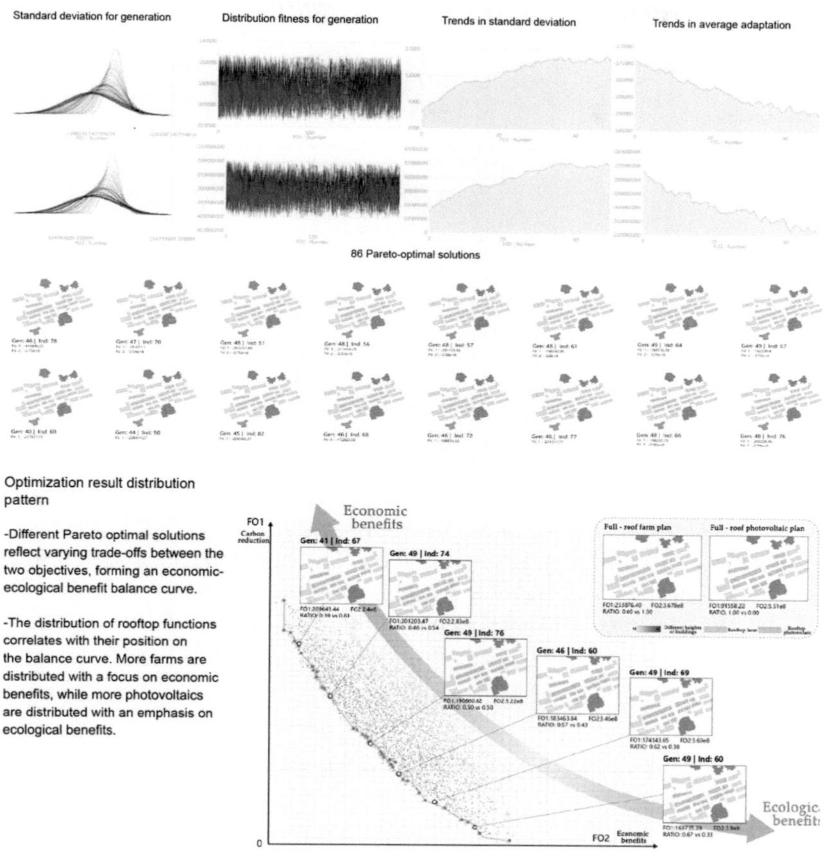

This section outlines the multi-objective optimization settings for the project, focusing on carbon reduction (FO1) and economic benefits (FO2). The optimization tool utilizes genetic algorithms with parameters like crossover and mutation probabilities, as well as population size, to produce optimized solutions. The analysis includes standard deviations and trends, showing 86 Pareto-optimal solutions that illustrate the trade-off between ecological and economic benefits. The results indicate that rooftop farms tend to offer more ecological benefits, while photovoltaic panels are linked to greater economic advantages. A graph visualizes this trade-off effectively.

11.8 Fitness Distribution of Optimization Results

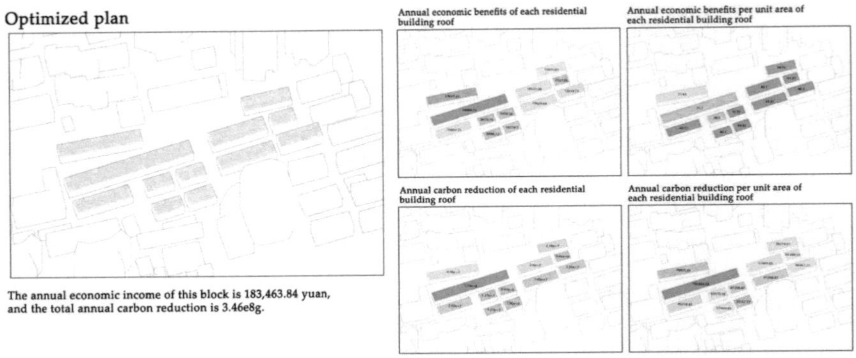

This section presents the results of the optimized plan and their distribution. The top-right image shows the annual economic benefits of each residential building's rooftop, with higher values indicating greater benefits. The bottom-right image displays annual carbon reduction per area for each rooftop. The top-left image highlights total annual economic benefits, while the bottom-left shows total annual carbon reduction for each rooftop.

11.9 Optimization Results of Roof with Different Module Sizes

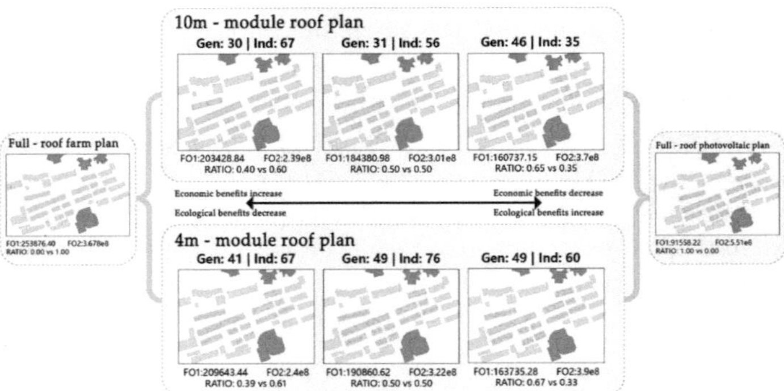

11.10 Effect of Mesh Size on Optimization Results

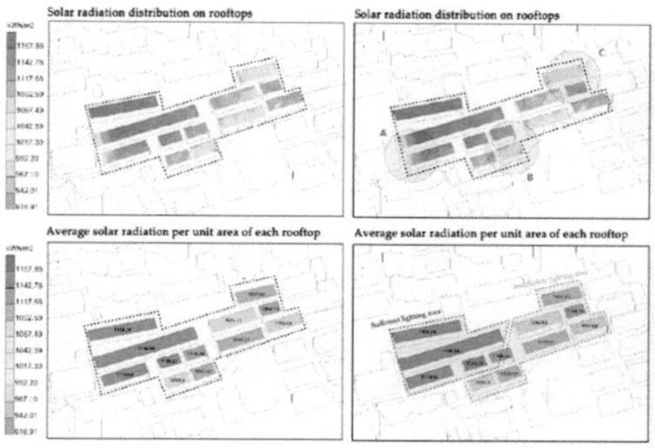

-Observing the overall distribution, photovoltaics are mostly in well-lit areas, while farms are mainly in less sunny spots. This is more evident in the 10-meter module scheme, showing distinct zoning. The 4-meter module scheme mixes functionalities based on this observation.

-Observing locations A, B, and C, areas with lower solar radiation due to building shading aren't suitable for photovoltaic panels. The 10-meter module scheme allocates larger farming areas to compensate, while the 4-meter module scheme maximizes solar resources with a targeted layout, resulting in higher benefits and carbon reduction.

11.11 Optimization Results of Other Sites

Xuhui District, Dong'an First Village

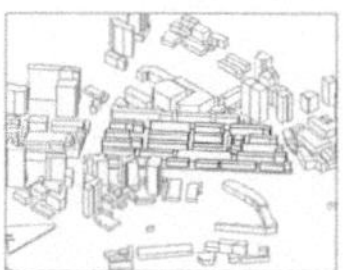

Selected target street area

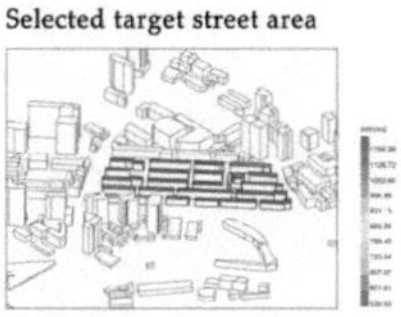

Sunlight simulation analysis

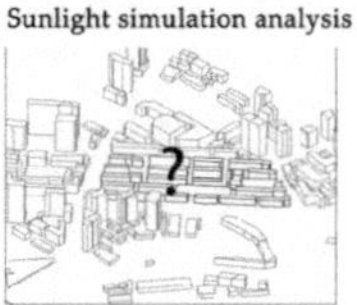

Roof performance optimization

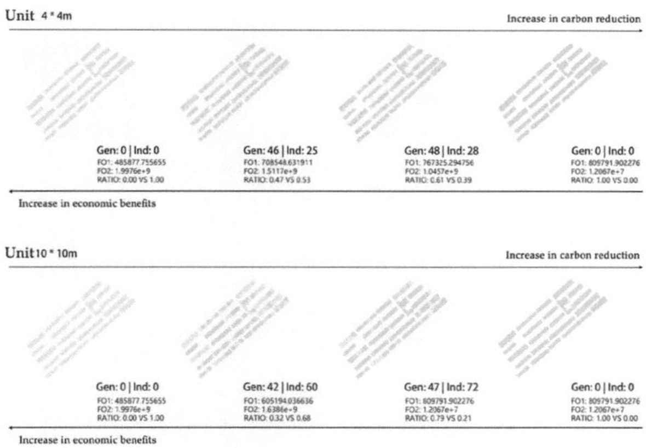

No. 120 Quxi Road, Huangpu District

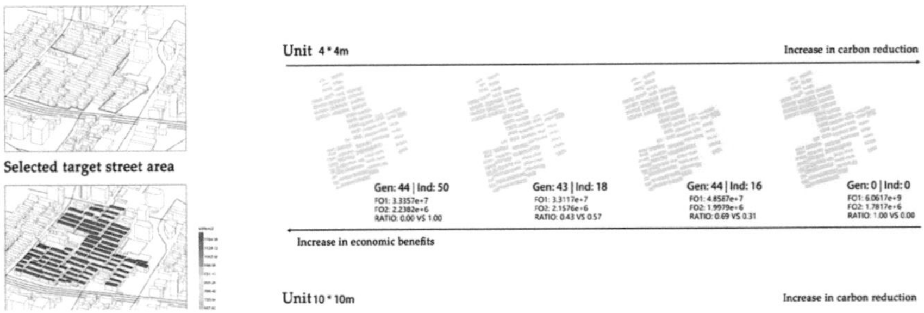

This section summarizes the optimization results from Dong'an First Village in Xuhui District and No. 120 Quxi Road in Huangpu District. By analyzing solar radiation and optimizing roof performance, the results compare economic benefits and carbon reduction for different grid sizes (4 × 4 m and 10 × 10 m). Smaller grids enhance the layout of photovoltaics and farming, yielding greater carbon reduction, while larger grids focus more on economic gains. The optimized plans aim to balance these objectives to meet the unique needs of each neighborhood.

No. 120 Quxi Road, Huangpu District

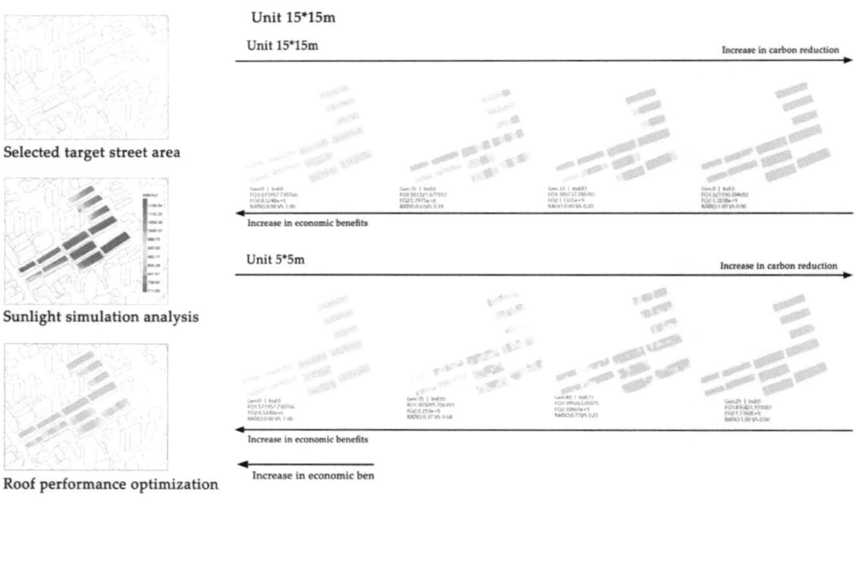

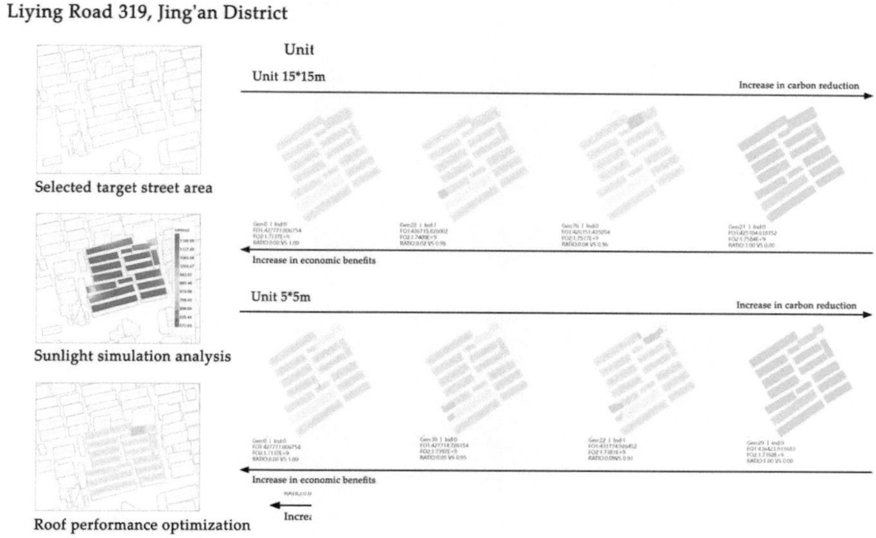

This section highlights the roof optimization results for two locations in Shanghai: No. 120 Quxi Road in Huangpu District and No. 319 Liying Road in Jing'an District. Solar radiation simulations were performed using grid sizes of 15 × 15 m and 5 × 5 m. The results show that smaller grids (5 × 5 m) provide greater carbon reduction and economic benefits in some areas, while larger grids (15 × 15 m) offer stronger economic advantages and a balanced approach to carbon reduction in others. By optimizing these grid schemes, we can enhance the roof functions, improving environmental benefits and economic returns with effective placement of photovoltaic and agricultural modules.

Chapter 12
CT-Marketing Low Carbon Community Design Based on Movable Modularity

Members

Chen Yutong, Jiang Qingrui, Li Danrui, Luo Zhaoyang, Zhang Hanqi

Background

Based on carbon neutrality goals, the national carbon emissions trading market is a major institutional innovation using market mechanisms to reduce greenhouse gas emissions and promote green development. Inspired by volumetric banking, there's a vision to create a future carbon sink bank.

Method

1. Carbon Bank Management

The planning bureau determines baseline carbon emissions for each plot based on its plot ratio, serving as the benchmark for carbon trading. Specific trading, compensation, and financial management are centrally coordinated by the government's carbon sink bank. Calculating potential carbon emissions for each plot becomes a crucial research aspect.

2. Carbon Simulation

J. Yao, *Intelligent Exploration in Sustainable Design*,
https://doi.org/10.1007/978-981-96-6743-7_12

Random sampling of architectural forms is conducted based on various plot ratios. Using a specific plot in Shanghai as an example, carbon energy assessments are performed on all possible building forms, resulting in carbon emission quotas. By averaging numerous simulations, potential carbon emissions for the plot are determined. Conclusion: Standard coal consumption per unit area varies for different plots under the same simulation conditions.

3. Building Optimization

Early-stage building design optimization is inspired by predicted potential carbon emissions for specific plots. An assembly-based architectural unit system is used to design buildings with climatic responsiveness in form to address different seasonal conditions. Machine learning employs appropriate algorithms (such as multilayer perceptrons, histogram gradient boosting, extreme random forests) to construct surrogate models for different datasets. Surrogate model technology, using RBM linear optimization and genetic algorithms, finds optimal solutions for building forms under different seasonal conditions.

Significance

A framework for operating a carbon sink bank in urban development has been designed, along with a proposed method for determining baseline carbon emissions for plots. Using these baseline emissions as a reference, a unit-style youth community apartment complex has been designed with climatic responsiveness in its overall form.

12.1 Technology Method

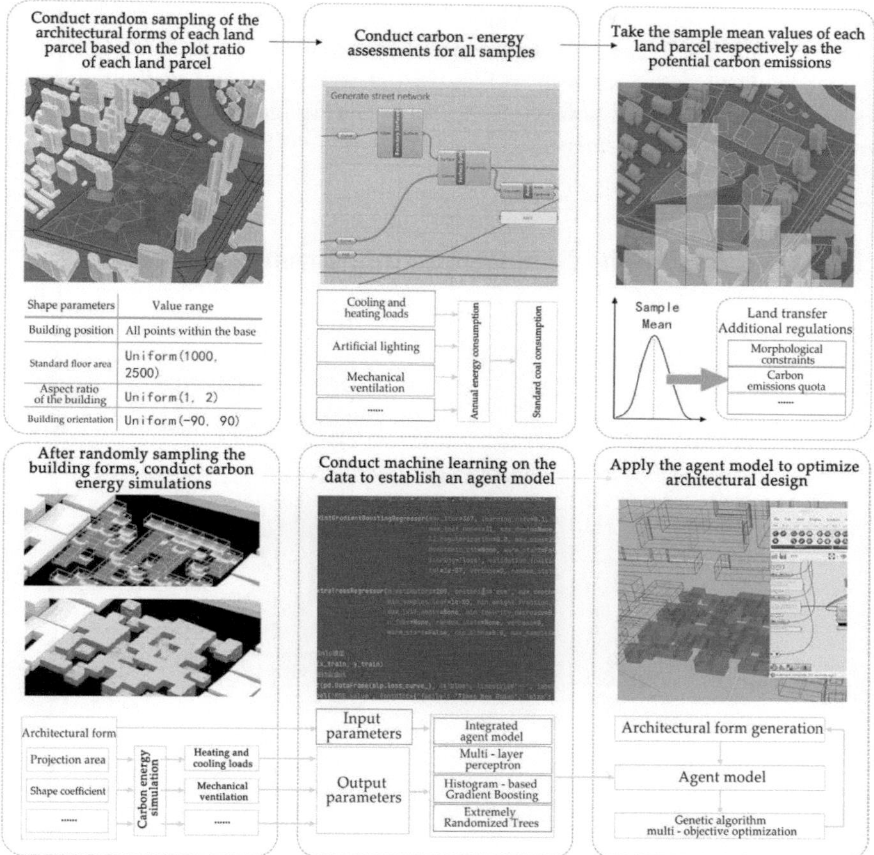

First, random sampling of architectural forms is carried out for each land parcel, followed by carbon-energy assessments that consider cooling and heating loads, lighting, and ventilation. The sample means determine potential carbon emissions, which are compared against land transfer and emission quotas. Next, machine learning is utilized to create an agent model for architectural design optimization through multi-objective algorithms like genetic algorithms and methods such as histogram-based gradient boosting. The optimized designs aim to reduce energy consumption and carbon emissions.

12.2 Carbon Credits Bank Regulatory Framework

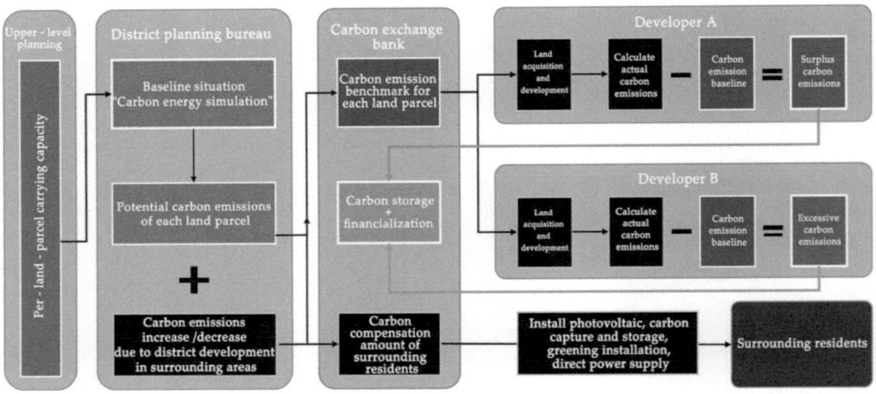

This text presents a framework for a carbon credits bank aimed at managing carbon emissions through planning and development. The district planning bureau first assesses the carbon emission potential of each land parcel by comparing the emissions with a baseline. If emissions exceed the baseline, surplus emissions are identified; if below, carbon reductions occur. Developers then calculate actual emissions during land acquisition and development. Based on these calculations, they can offset excess emissions or enhance carbon storage through measures like photovoltaic panels and carbon capture systems. This framework fosters carbon offset trading and supports green infrastructure development.

12.3 Simulation Results

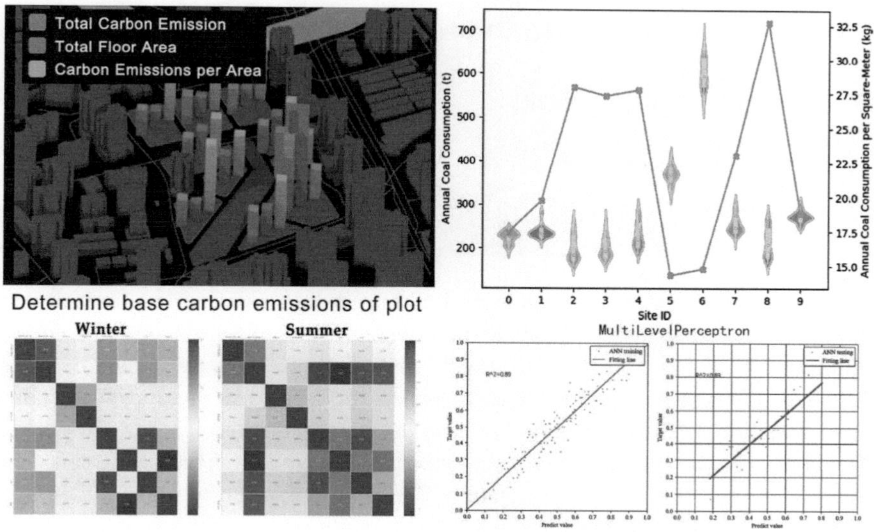

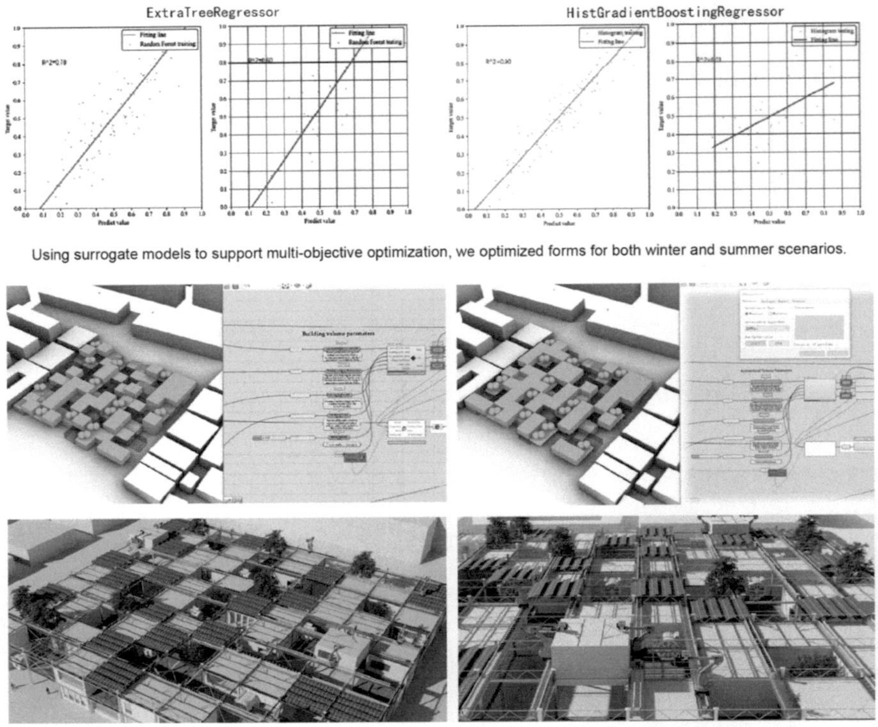

This section outlines how to optimize building design using simulation results. Carbon emissions and energy consumption were simulated for winter and summer conditions. Utilizing agent-based models and machine learning techniques like Multi-Level Perceptron and ExtraTreeRegressor, the design team predicted and optimized carbon emissions based on building parameters derived from training data. The team integrated building massing, layout, and energy efficiency metrics, using Grasshopper and Rhino for visualization. The simulations led to improved energy efficiency and reduced carbon emissions. Additionally, robotic technology was incorporated in the construction process, including the installation of solar panels and the use of green materials, promoting a more sustainable building design.

Chapter 13
Urban Carbon Emission Predictions in Response to Demographic Change

Members
Cui Qinyu, Zhang Chi, Feng Jiajia, Bi Xuejiao, Wang Lei

Background
Under the 2060 carbon neutrality goal, cutting carbon emissions is urgent, especially as the aging population grows. In our country, people typically retire at 60, leading to less physical activity and higher energy use at home. This situation affects the energy needed to run buildings and the carbon emissions in cities. This research looks into how an aging population impacts city carbon emissions.

Method
1. Scope Selection

This study targets the central area of Guangzhou, which holds about 80% of the city's main population. Using software and data analysis methods, around 570,000 buildings were identified, focusing on residential, office, and commercial buildings. Residential buildings make up 86% of this total, while office buildings are about 1%.

2. Population Data Analysis

The team analyzed 5.5 million data points from smart footprint logistics, capturing details like age, gender, and living or working locations with high accuracy. A linear regression was used to understand how the elderly population affects building occupancy. Findings show residential occupancy increases with a higher elderly population, while office and commercial occupancies decrease.

J. Yao, *Intelligent Exploration in Sustainable Design*,
https://doi.org/10.1007/978-981-96-6743-7_13

3. Predicting Occupancy Rates

The study predicts changes in building occupancy from 2020 to 2050 based on the aging population. It's expected that residential occupancy will rise by 15.9%, while office occupancy will drop by 18.6%, and commercial by 7.3%.

4. Carbon Emission Predictions

The research looks at how building occupancy rates influence energy use and, consequently, carbon emissions. Residential buildings are most affected (75% by occupancy rates), while offices and commercial buildings are less so (20%). By considering these factors and the aging population's growth, the study forecasts urban carbon emissions from 2020 to 2050, showing a direct link between aging and emissions.

Significance

The study finds that by 2050, the elderly population in Guangzhou's central area is expected to increase from 18% to 38%, causing a 3.1% rise in carbon emissions (about 997,000 tons/year) from its buildings. Residential buildings will see a notable increase in emissions, up by 10.3% (around 1,371,000 tons/year). With aging population trends, future building strategies should focus more on saving energy and adding carbon offset measures in residential zones.

13.1 Technology Method

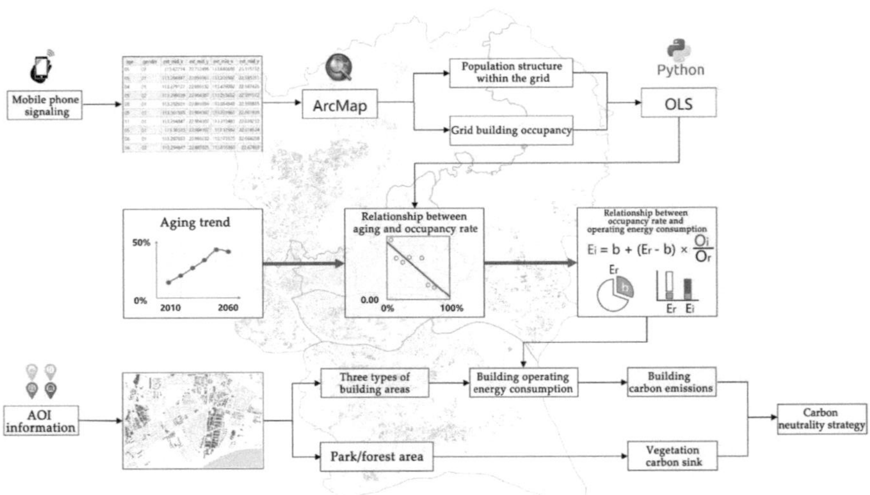

This part outlines a building energy efficiency analysis method using big data and machine learning. Mobile phone signal data is utilized to gather population information for occupancy analysis via ArcMap. By modeling the relationship between population aging trends and building occupancy rates, energy consumption predictions are made using the Ordinary Least Squares (OLS) method. The model assesses energy use in different areas, factoring in green spaces to evaluate carbon emissions and develop a carbon neutrality strategy. Additionally, integrating local facility information helps optimize building carbon emissions and energy efficiency.

13.2 Research Scope

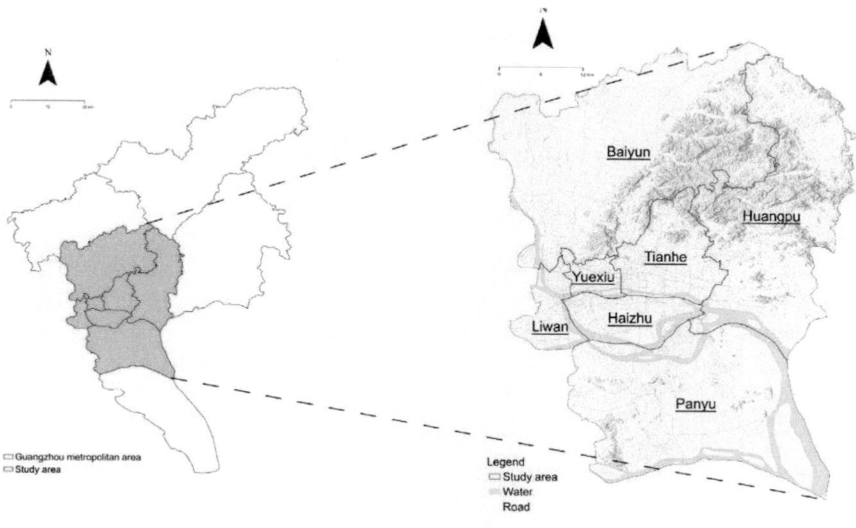

This section shows the geographical scope of the study, which is mainly focused on the urban areas of Guangzhou.

13.3 Research Data

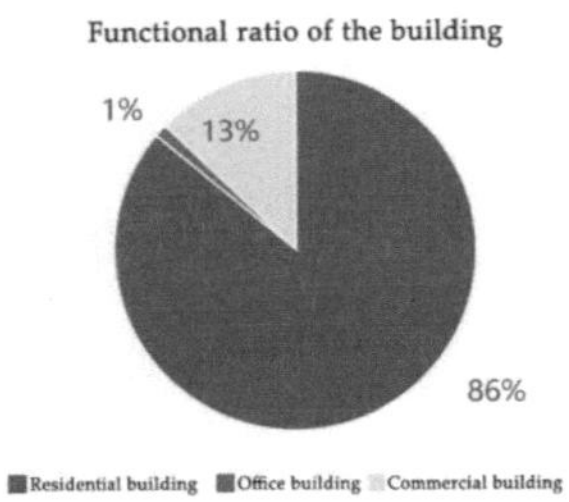

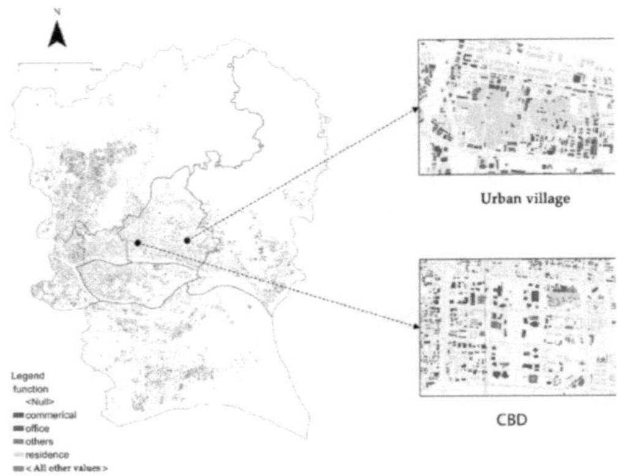

This study examines the functional distribution of buildings in the area, revealing that 86% are residential, while office and commercial buildings comprise a smaller fraction. Maps illustrate the zoning of these functions in an urban village and central business district (CBD).

13.4 Research Method

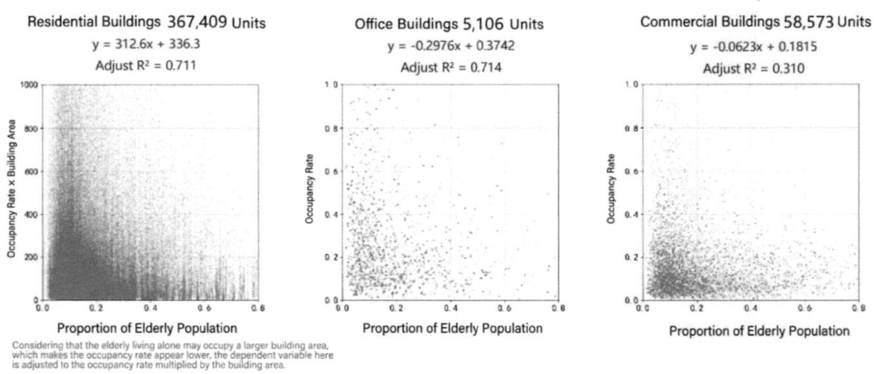

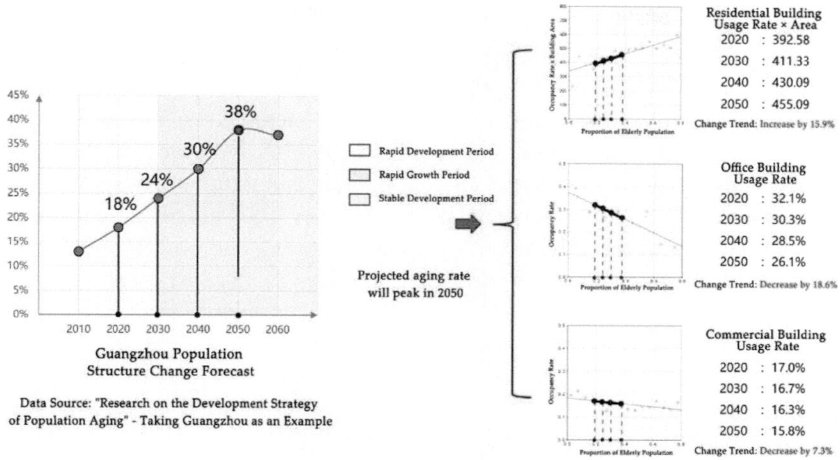

Fitting the relationship between building vacancy rates and changes in population ageing rates

This section analyzes the relationship between building vacancy rates and population aging. It examines the correlation between the elderly population and vacancy rates in residential, office, and commercial buildings. For residential buildings, a higher aging population correlates with increased vacancy rates, while for office and commercial buildings, it correlates negatively. Using Guangzhou's aging population forecast, the model predicts the aging rate will peak by 2050 and its impact on building occupancy rates.

13.5 Data Comparison

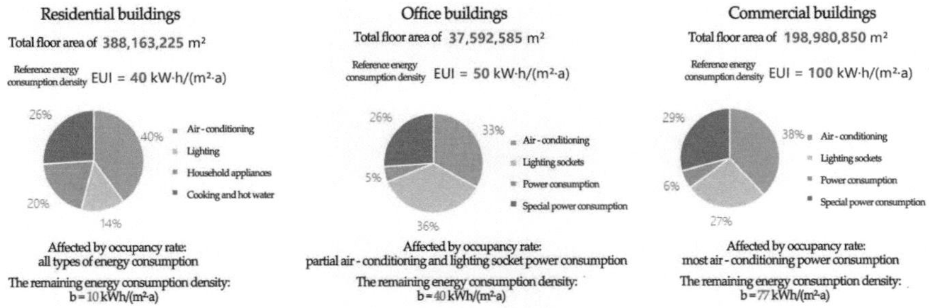

Estimating carbon emissions from different types of buildings based on occupancy rates

This section estimates carbon emissions from various building types based on occupancy rates. For residential buildings, with a total floor area of 388,163,225 m^2, the energy use intensity (EUI) is 40 kWh/(m^2·a), affected by occupancy-related demands like air conditioning and lighting, with a remaining energy consumption density of 10 kWh/(m^2·a). Office buildings, totaling 37,592,585 m^2, have an EUI of 50 kWh/(m^2·a), mainly influenced by partial air conditioning and lighting, with a remaining density of 40 kWh/(m^2·a). Commercial buildings cover 198,980,850 m^2, with an EUI of 100 kWh/(m^2·a), primarily driven by air conditioning and power consumption, and a remaining density of 77 kWh/(m^2·a). Carbon emissions are estimated for each building type based on their energy consumption and occupancy rates.

13.6 Research Results

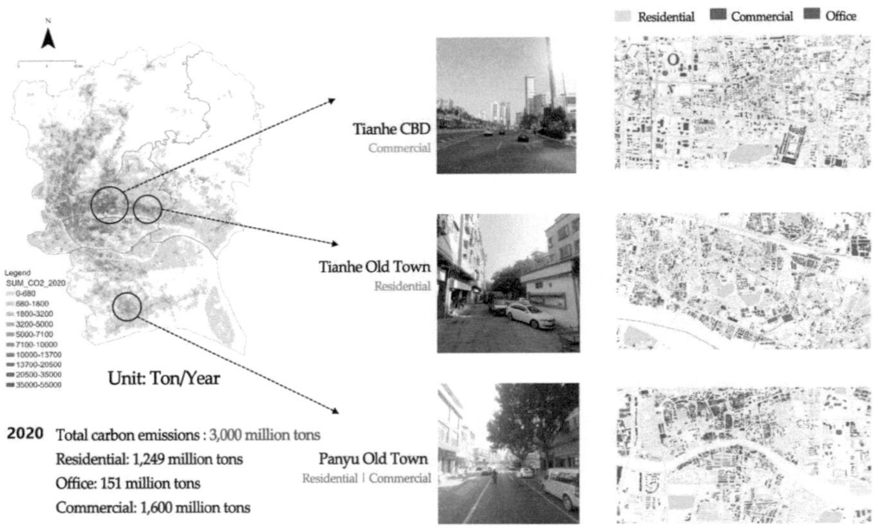

Estimated carbon emissions from buildings in Guangzhou in 2020

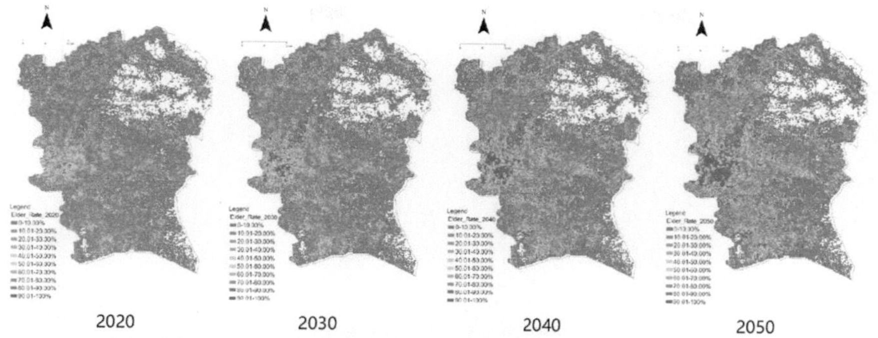

Extrapolation of spatial and temporal distribution of aging rate in downtown Guangzhou, 2020-2050

This section presents the estimated carbon emissions from buildings in Guangzhou in 2020. Total emissions reached 3000 million tons, with residential buildings contributing 1249 million tons, office buildings 151 million tons, and commercial buildings 1600 million tons. The images showcase areas like Tianhe CBD, Tianhe Old Town, and Panyu Old Town along with their building types. Maps illustrate the aging rates in downtown Guangzhou from 2020 to 2050, showing how the aging population will affect energy consumption and carbon emissions. The charts indicate an increasing aging rate in certain areas, which will impact building performance and carbon footprint in the future.

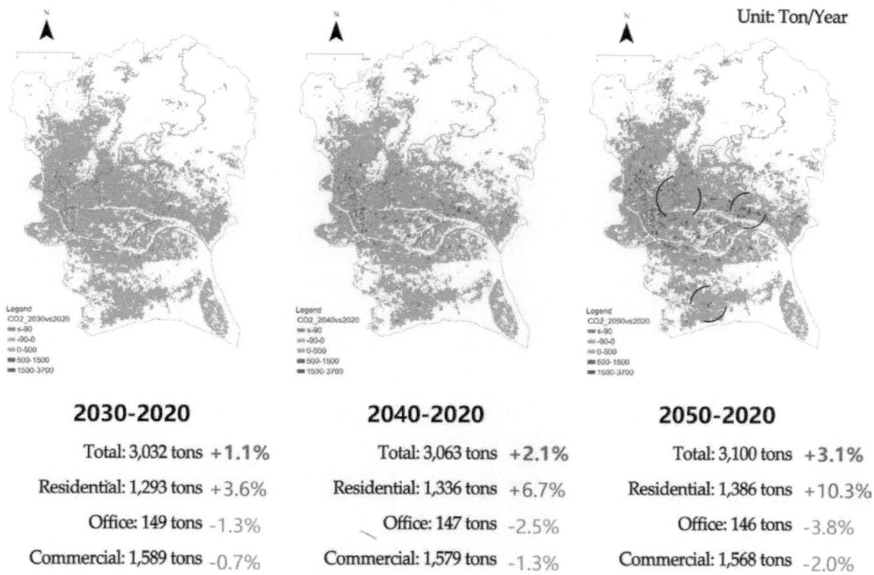

Carbon emission trends in Guangzhou downtown area, 2030–2050

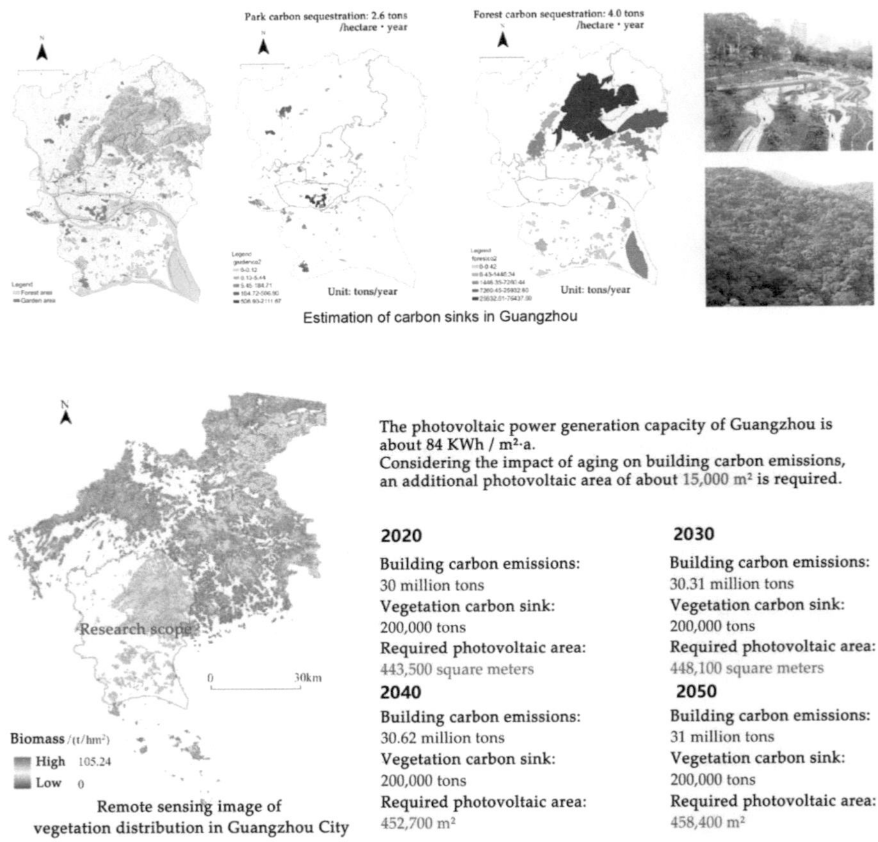

Carbon neutral strategies

This section outlines carbon emission trends and sequestration estimates for Guangzhou from 2020 to 2050. It highlights a projected increase in emissions driven by population aging, with residential, office, and commercial sectors showing a rise of 1.1% by 2030 and 3.1% by 2050. Additional photovoltaic installations are necessary to offset these emissions, and carbon sequestration from park and forest areas is estimated at 2.6 tons and 4.0 tons per hectare per year, respectively. To achieve carbon neutrality, significant efforts in installing photovoltaic panels and enhancing green infrastructure are crucial. The strategy integrates renewable energy measures with urban planning to meet future emissions targets.

Chapter 14
Urban Carbon Neutrality Prediction and Optimization: A Case Study of the Integrated Demonstration Zone in the Yangtze River Delta

Members
Lu Yijun, Yu Mengqing, Chen Qiyue, Wu Zihao, Yu Bingqing

Background
As global warming and urban heat issues worsen, and China's urbanization speeds up, the environmental impact of energy consumption is set to increase. This highlights the need to manage carbon emissions better. The research aims to develop a simplified platform that visualizes land use and carbon data, allowing for the analysis of urban carbon based on different policy. This tool is intended to help with better environmental planning and carbon management in cities.

Method
1. Building the Carbon Model

For this study, the Integrated Development Demonstration Zone in the Yangtze River Delta was chosen. Using GIS land use and land cover (LULC) data collection, we incorporated carbon emission coefficients into the GIS system. This, coupled with existing land types, formed a bottom-up urban carbon model.

2. Carbon Scenario Prediction

Building upon the carbon model, we employed the Geographic Simulation System GeoSOS-FLUS to predict land use evolution. Initially, an Artificial Neural Network

J. Yao, *Intelligent Exploration in Sustainable Design*,
https://doi.org/10.1007/978-981-96-6743-7_14

(ANN) algorithm determined suitability probabilities for various land uses. Subsequently, urban growth was simulated using a combination of System Dynamics (SD) and Cellular Automata (CA) models, addressing land use conversion under both natural and human influences.

3. Comparison of Future Development

We established five simulated scenarios of land use change: baseline, industrial development, urban development, coordinated urban and green space development, and ecological protection. Each scenario exhibited varying impacts on carbon reduction. Notably, the ecological protection scenario demonstrated significant improvements in carbon sink capacity despite persistent carbon deficits.

4. Impact on LST

Landsat 8 images were obtained and processed to derive land surface temperature maps. Analysis revealed a correlation between land use types and surface temperature, as demonstrated by a regression equation with an R-square value of 0.709.

Significance
Urban planning contributes to carbon neutrality by influencing land use, which involves continuously changing carbon emissions. Future discussions may focus on carbon emission coefficients across different time sequences and predict their values to reflect energy usage under optimized energy structures in the long term.

14.1 Technology Method

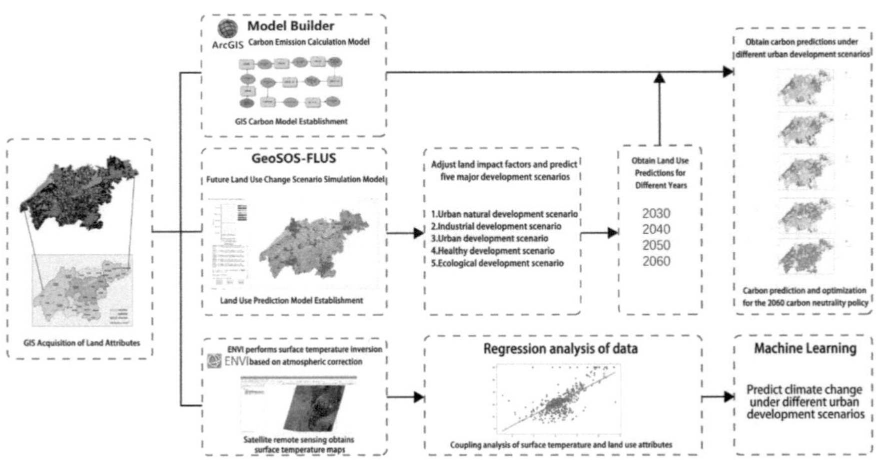

This section outlines the methodology for predicting carbon emissions and land use changes under various urban development scenarios using GIS, machine learning, and land use simulation models. It starts with GIS to gather land attributes, which are input into the GeoSOS-FLUS model to forecast land use changes for 2030, 2040, 2050, and 2060 based on five development scenarios: urban natural, industrial, urban, healthy, and ecological development. The model then employs ENVI software for surface temperature inversion, analyzing satellite data to create temperature maps. Regression analysis of these maps alongside land use data refines carbon emissions predictions for each scenario. Finally, machine learning is used to project climate change impacts based on the urban development strategies, aiming to inform pathways to achieve carbon neutrality goals by 2060.

14.2 Land Use Data

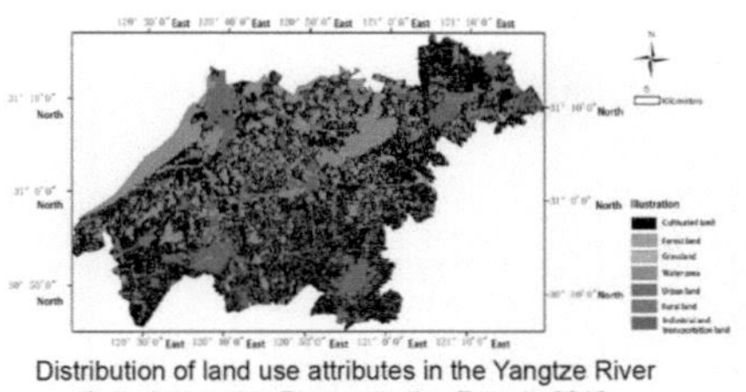

Distribution of land use attributes in the Yangtze River Delta Integration Demonstration Zone in 2010

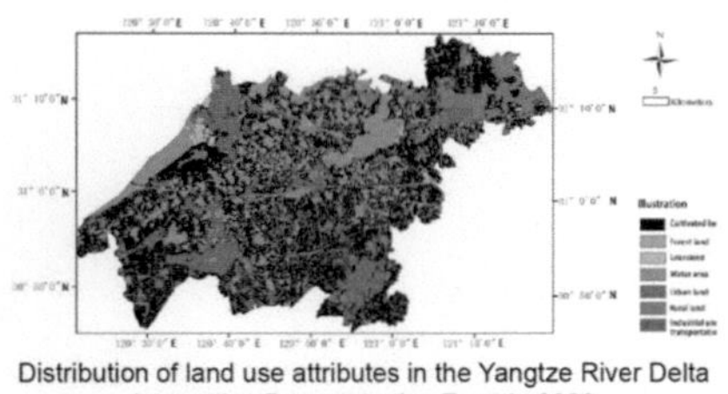

Distribution of land use attributes in the Yangtze River Delta Integration Demonstration Zone in 2020

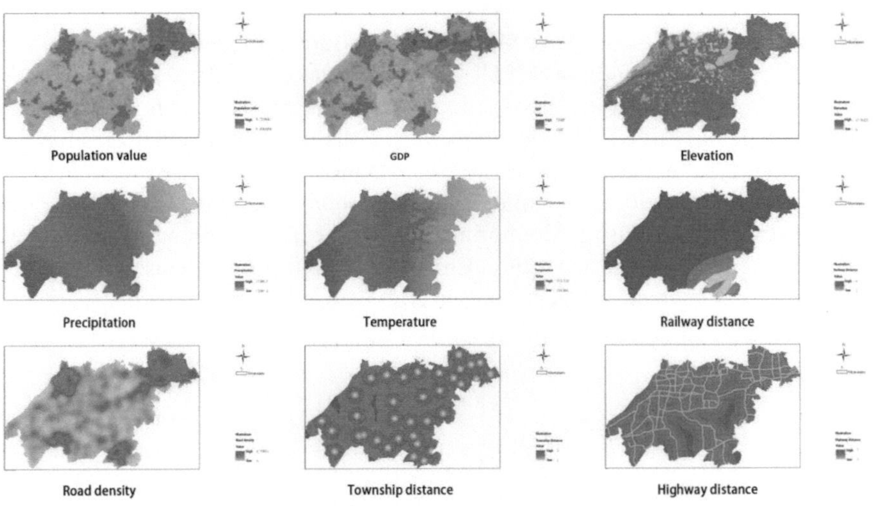

This section presents land use data for the Yangtze River Delta Integration Demonstration Zone, highlighting changes in land types like cultivated land, built-up land, forest, and wetland from 2010 to 2020. It includes indicators such as population density, GDP, elevation, precipitation, temperature, road density, and distances to townships and highways, all shown through maps. This information aids in analyzing economic activities, environmental conditions, and transportation networks for better land use planning and carbon emission predictions.

14.3 Land-Use Oriented Urban Carbon Modeling

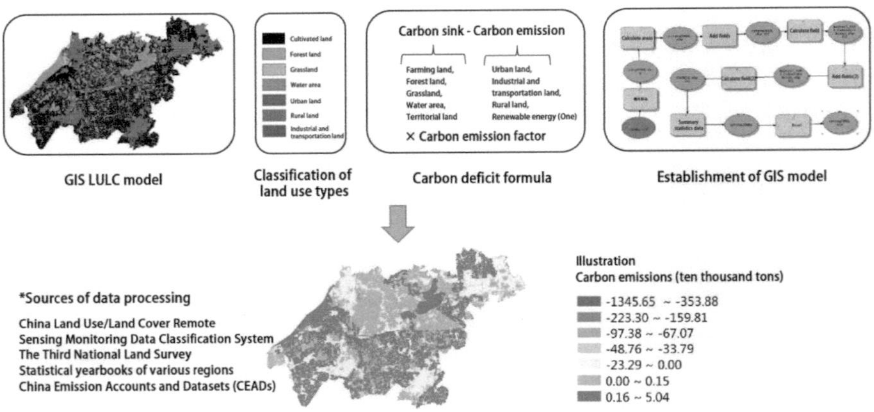

This section highlights a Land-use Oriented Urban Carbon Modeling approach using the GIS Land Use and Land Cover (LULC) model to classify land types like cultivated land, forest, grassland, water areas, urban, and industrial land. This classification aids in calculating carbon sinks and emissions, utilizing a formula that accounts for different land types and emission factors. The GIS model involves steps such as calculating areas, adding fields, and summarizing data, sourced from the China Land Use/ Land Cover Remote Sensing Monitoring System, the Third National Land Survey, Statistical Yearbooks, and the China Emission Accounts and Datasets (CEADs). The final model estimates regional carbon emissions, visualized on a map showing varying emission levels.

14.4 Model Parameters

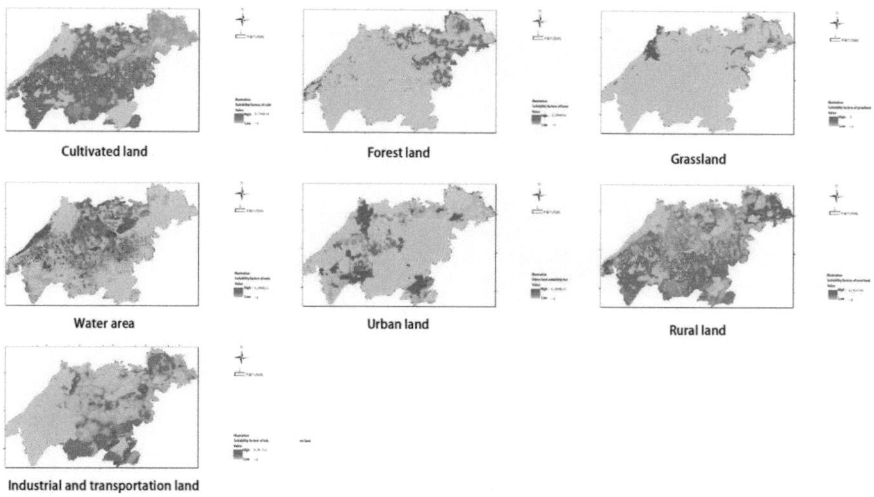

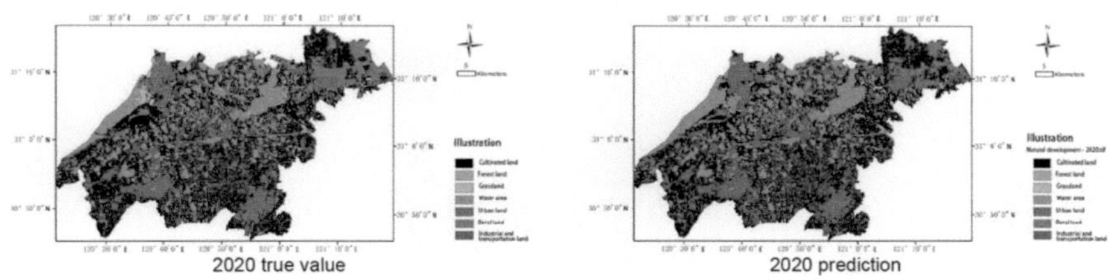

This image shows the model's input data, featuring the distribution of land types (like cultivated land, forest, and grassland) in the region. It compares the true land type distribution in 2020 with the model's predicted values, allowing evaluation of the model's accuracy in predicting land use types.

14.5 Multi-scenarios

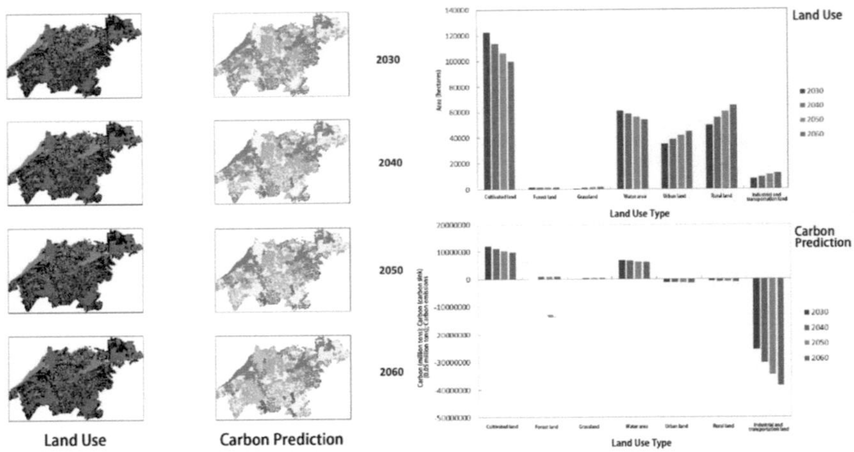

Base scenario

The baseline scenario projects future developments based on current trends without interventions. It analyzes changes in land use (like cultivated land, forests, and urban areas) and their impact on carbon emissions and sinks.

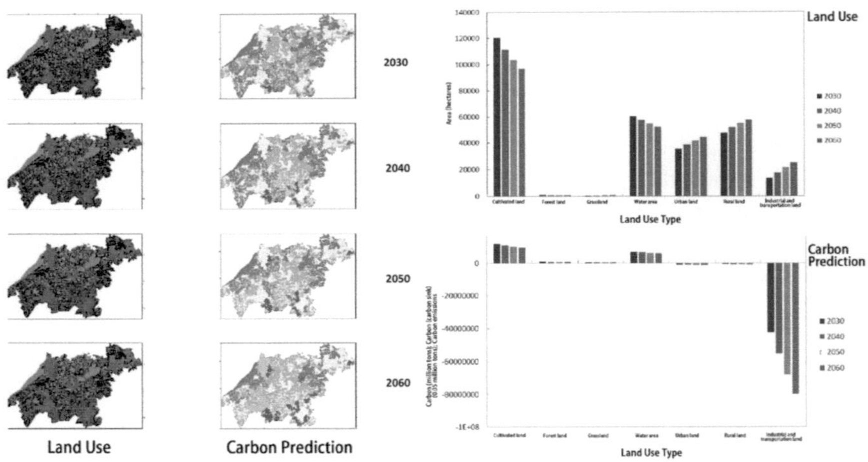

Industrial development scenario

The industrial development scenario describes a shift in land use towards industry, transportation, and urban construction amid rapid economic growth. It highlights trends in land use changes and their impact on carbon emissions for the years 2030, 2040, 2050, and 2060. As industrial and transportation areas expand, carbon emissions increase significantly, carbon sink capacity decreases, and environmental pressures rise.

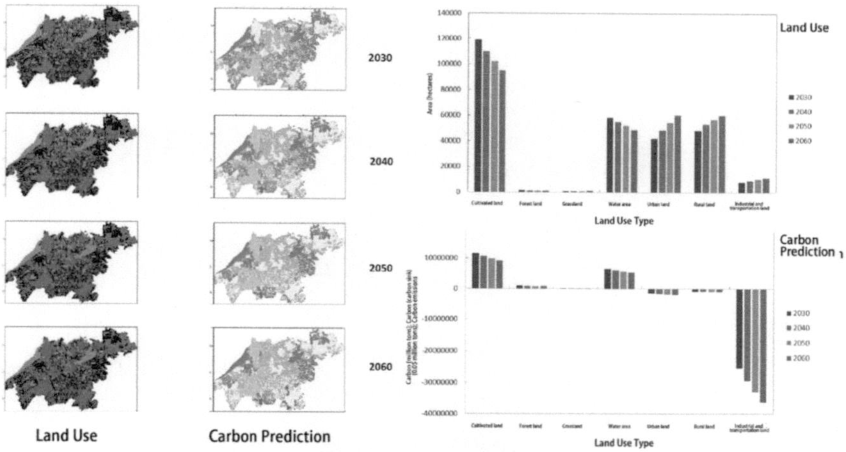

Urban development scenario

The baseline scenario projects future developments based on current trends without interventions. It analyzes changes in land use (like cultivated land, forests, and urban areas) and their impact on carbon emissions and sinks.

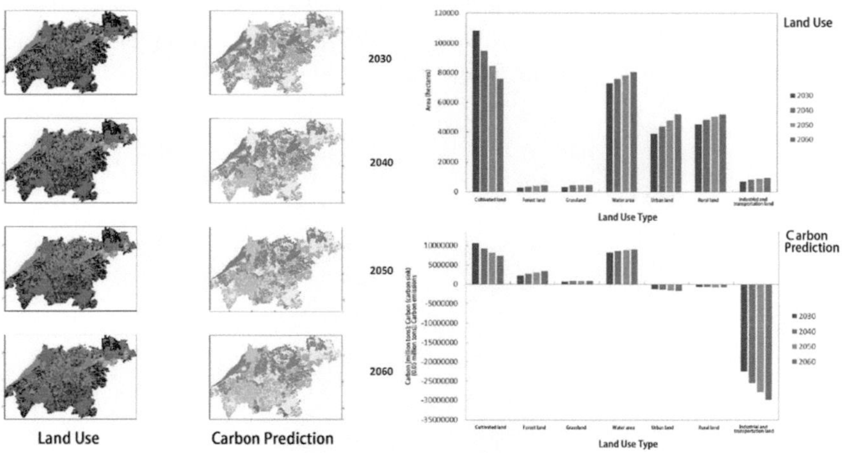

Healthy urban development scenario

The Healthy Urban Development Scenario balances urbanization with ecological protection and sustainable development. It emphasizes rational land use planning to minimize environmental impact. The scenario shows controlled urban expansion, preservation of forests and green spaces, and a relatively smaller increase in carbon emissions, enhancing carbon sink capacity and promoting sustainability.

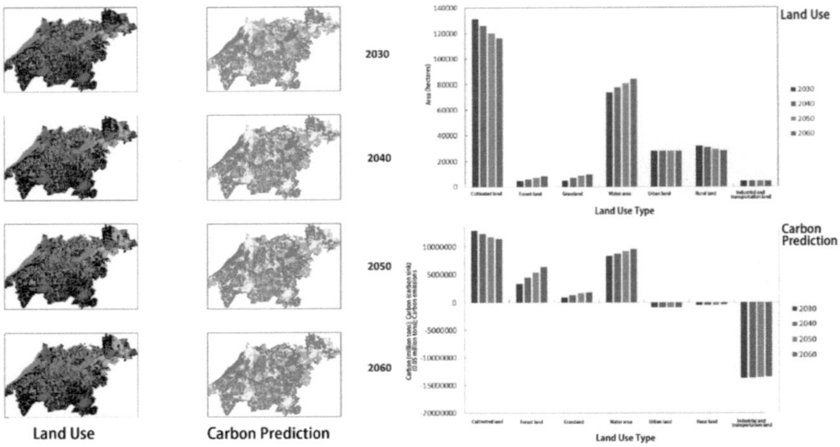

Ecological conservation scenario

The Ecological Conservation Scenario focuses on prioritizing environmental protection and sustainability by limiting land development and preserving natural areas. It illustrates land use changes and their effects on carbon emissions for 2030, 2040, 2050, and 2060, showing that cultivated land and ecological areas remain protected, while forest and grassland areas increase. This results in enhanced carbon sink capacity and significantly reduced carbon emissions, aiding environmental improvement and climate change mitigation.

14.6 Scenarios Comparisons

Scenario Type	Urban Land Area	Industrial and Transportation Land Area
Baseline Scenario	100%	100%
Industrial Development Scenario	99.52%	208.19%
Urban Development Scenario	135.31%	94.19%
Healthy Development Scenario	116.28%	77.56%
Ecological Protection Scenario	62.27%	35.54%

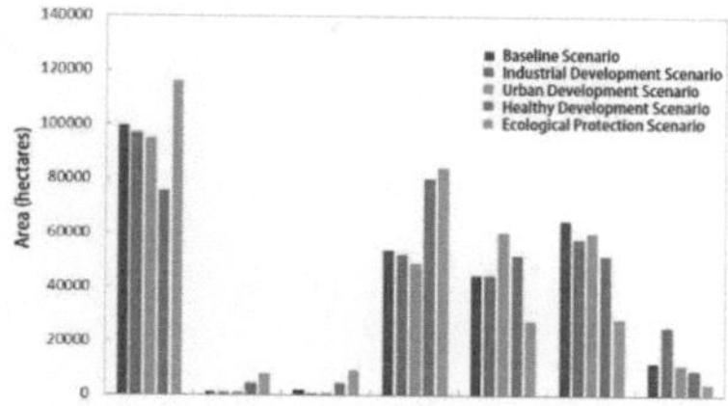

Scenario Type	Urban Land Area	Carbon Emission
Baseline Scenario	100%	100%
Industrial Development Scenario	94.03%	201.31%
Urban Development Scenario	92.69%	95.80%
Healthy Development Scenario	121.53%	79.15%
Ecological Protection Scenario	169.63%	36.14%

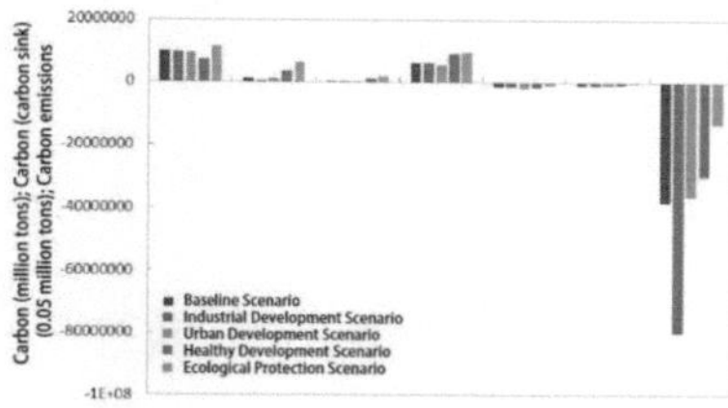

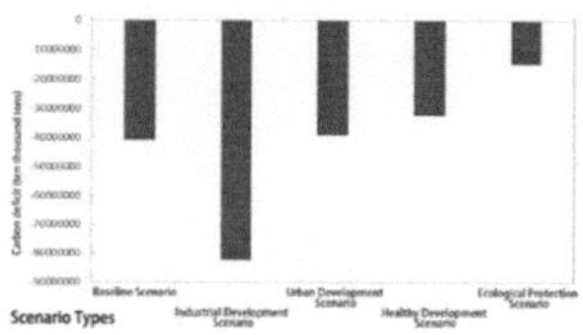

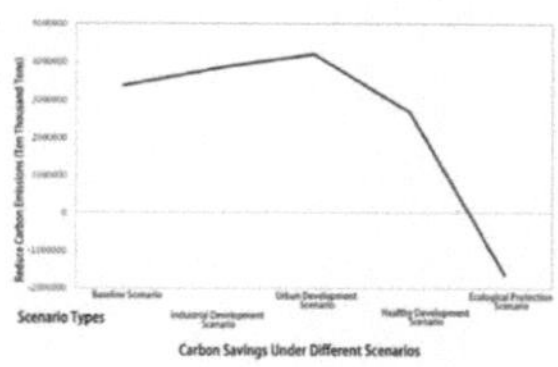

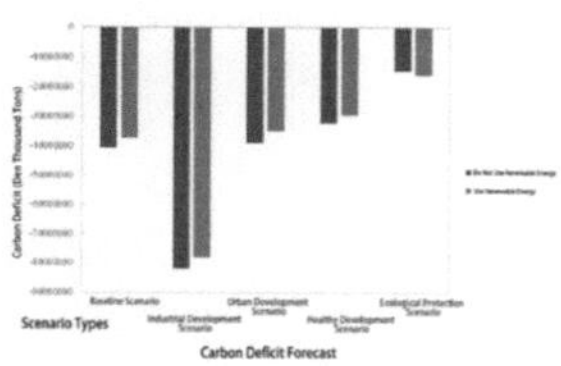

Optimization strategy

This figure analyzes various land use scenarios (Baseline, Industrial Development, Urban Development, Healthy Development, and Ecological Protection) regarding urban and industrial land changes and their impact on carbon emissions. The Industrial Development Scenario leads to significant industrial land increase and the highest carbon emissions, while the Ecological Protection Scenario restricts urban expansion, resulting in the lowest emissions. The analysis indicates that the Healthy Development and Ecological Protection Scenarios optimize carbon balance, supporting sustainable land use planning and carbon reduction policies.

Chapter 15
Analysis and Optimization of Urban Zero-Carbon Mobility Layout from the Perspective of Bike-Sharing

Members
Wen Wen, Zhang Yecheng, Qi Ziying, Xiong Hongyi, Wang Zixuan, Xu Wenya

Background
In recent years, bike-sharing has become a popular and eco-friendly way to travel in many cities, playing a key role in solving the 'last mile' problem in urban public transport by replacing the energy used by other modes of transport. However, bike-sharing faces issues such as low usage and turnover rates, poor spatial layout, and delayed resource allocation. This study aims to explore how bike-sharing usage patterns relate to urban spaces, analyzing the link between zero-carbon commuting and urban layouts to offer insights for better spatial planning of bike-sharing resources, supporting eco-friendly transportation.

Method
1. Data Collection and Preprocessing

Utilizing data from November 2021, we analyzed 1,046,794 bike-sharing GPS trip records in Wuhan to extract basic information on usage frequency, trip distances, and departure times. Morning and evening peak hours were identified as representative periods for in-depth analysis.

2. Regression

Considering various impacts of different commercial and public sectors on bike-sharing, the analysis incorporated POI data across nine broad categories, including dining, shopping, lodging, and educational facilities. The method involved using GIS to spatially distribute data and assess it through five key variables—accessibility, permeability, POI density, population, and land-use diversity—to measure spatial correlation. A comparison between OLS and GWR methods showed that GWR had superior fitting capabilities, revealing significant spatial autocorrelation among the variables.

J. Yao, *Intelligent Exploration in Sustainable Design*,
https://doi.org/10.1007/978-981-96-6743-7_15

3. Cluster and Prediction

Using machine learning (XGBoost), we built a model from 80% of our data to predict where bike-sharing trips start and end. Tested with the remaining 20%, this model showed high accuracy in predicting locations.

Significance
Research results show a significant correlation between urban population density, POI density, land use mix, urban road accessibility, and proximity with shared bicycle travel patterns. Using unsupervised learning, existing urban shared bicycle trajectories can be categorized into 5 groups. By adjusting urban spatial factors at the kilometer grid scale, a predictive model for bicycle mode choice can be developed, laying the groundwork for future carbon reduction research focused on public transit.

15.1 Technology Method

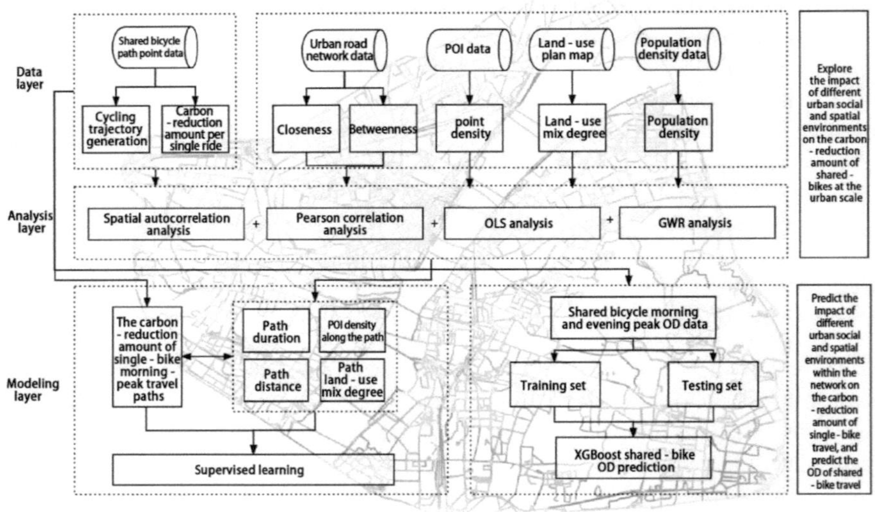

This technology roadmap outlines a research methodology for evaluating the carbon reduction impact of shared bicycle travel, organized into three layers: data, analysis, and modeling. The data layer consists of shared bicycle path data, urban road networks, points of interest (POIs), land-use information, and population density, creating a comprehensive view of the urban environment. The analysis layer utilizes spatial autocorrelation, Pearson correlation, Ordinary Least Squares (OLS), and Geographically Weighted Regression (GWR) to examine how various urban factors affect the carbon reduction potential of shared bicycles. The modeling layer employs supervised learning techniques, factoring in path duration, distance, POI density, and land-use mix to estimate the carbon reduction from individual trips. The XGBoost model further predicts origin-destination patterns of shared bicycle usage during peak hours. Ultimately, this roadmap aims to assess the carbon reduction potential of shared bicycles at an urban level and forecast travel patterns based on different environmental conditions, aiding sustainable urban mobility and carbon management efforts.

15.2 Analysis of Bicycle Sharing Usage in Wuhan

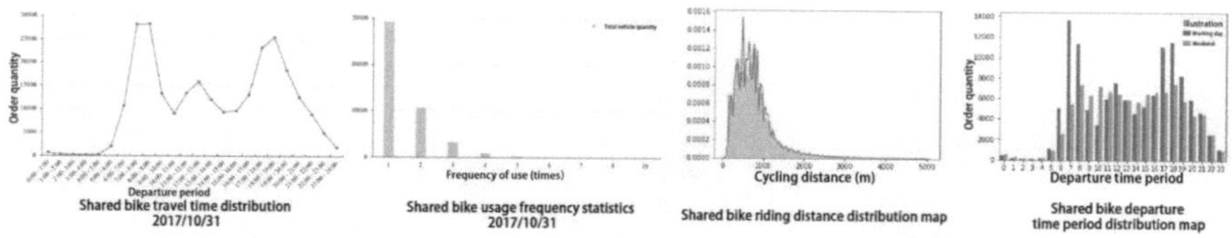

This analysis of bicycle-sharing in Wuhan covers travel time, usage frequency, cycling distance, and departure times. The first chart reveals peak travel times during morning and evening commutes. The second shows that most users ride only once, indicating short-term usage predominates. The third chart highlights that most trips are short, typically a few hundred meters to a few kilometers. Finally, weekday usage features more pronounced peaks, while weekend usage is more balanced. This insight is valuable for urban transportation planning and optimizing bike-sharing systems.

15.3 Results of sDNA Analysis

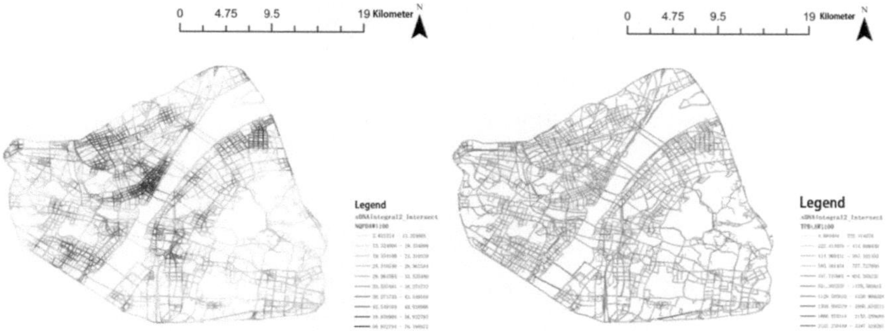

This figure shows the results of sDNA (spatial Design Network Analysis) to evaluate urban road network accessibility and connectivity. The left map uses a heatmap to indicate accessibility, with darker colors representing higher accessibility in areas like city centers. The right map features blue lines to illustrate road connectivity, with varying shades indicating the importance of different routes. sDNA analysis aids in understanding urban road structure, identifying high-traffic areas, and supporting urban planning and transportation optimization, including shared bicycle station placement.

15.4 Regression Analysis

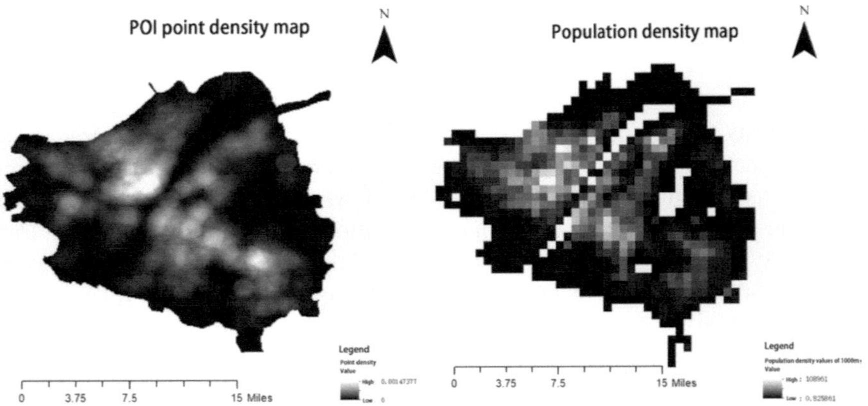

This figure displays two key variables in the regression analysis: POI (Point of Interest) density and population density. The left map shows POI density, with white areas indicating high concentrations in commercial centers and transportation hubs. The right map illustrates population density, where white regions represent densely populated residential or city core areas. These variables help analyze the relationship between urban characteristics, shared bike usage, and carbon reduction, aiding in bike station placement and transportation optimization.

15.5 Data Collection

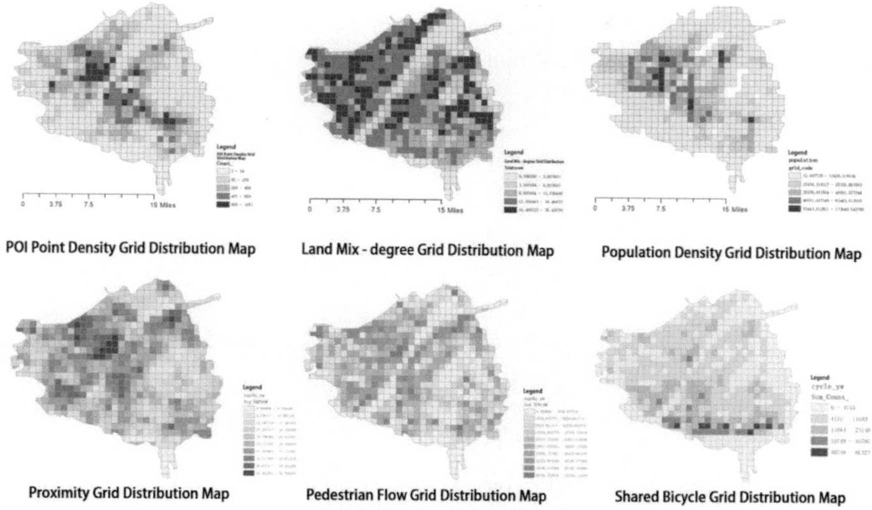

This figure shows the grid distribution of urban data such as POI density, land-use mix, population density, closeness, pedestrian flow, and shared bike usage. These metrics help analyze urban characteristics and their effects on shared bike travel and carbon reduction, aiding transportation planning and optimizing bike-sharing systems.

15.6 Spatial Autocorrelation Analysis

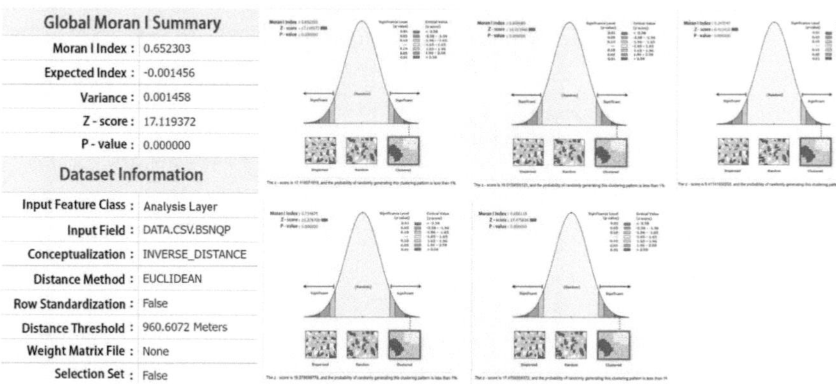

Global Moran I Summary	
Moran I Index :	0.652303
Expected Index :	-0.001456
Variance :	0.001458
Z - score :	17.119372
P - value :	0.000000
Dataset Information	
Input Feature Class :	Analysis Layer
Input Field :	DATA.CSV.BSNQP
Conceptualization :	INVERSE_DISTANCE
Distance Method :	EUCLIDEAN
Row Standardization :	False
Distance Threshold :	960.6072 Meters
Weight Matrix File :	None
Selection Set :	False

This figure shows the results of the Spatial Autocorrelation Analysis (Global Moran's I), revealing significant clustering in the data's spatial distribution. The Moran's I index is 0.652303, with a Z-score of 17.119372 and a near-zero P-value, indicating a strong clustering effect. The charts illustrate the Moran's I indices and distribution curves for various datasets, all demonstrating spatial clustering.

15.7 Pearson Correlation

pearson	value
ANQ	0.480**
ATPB	0.303**
Poi	0.534**
People	0.549**
Mixed	0.687**

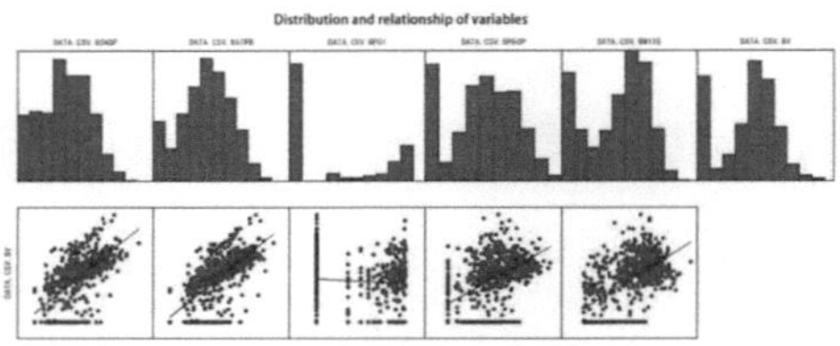

The figure presents the Pearson correlation analysis results, showing correlation coefficients for various variables. The "Mixed" variable has the highest correlation (0.687), while "ATPB" has the lowest (0.303). The right charts display variable distributions and scatter plots, indicating some linear correlation among the variables.

15.8 OLS and GWR Analysis

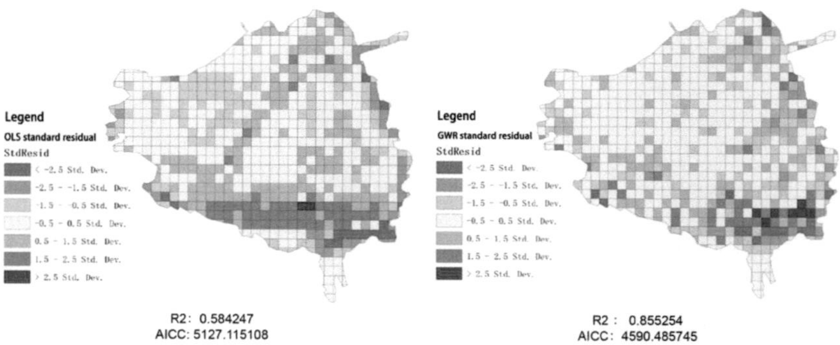

This figure compares the results of Ordinary Least Squares (OLS) regression and Geographically Weighted Regression (GWR). The left map shows the OLS standard residual distribution, while the right map displays that of GWR. The GWR model has a higher R2 value (0.855254) and a lower AICC value (4590.485745 vs. 5127.115108), indicating better spatial interpretation and fit. The colors represent the standard deviation of residuals, with red areas showing positive residuals and blue areas indicating negative residuals, highlighting the GWR model's ability to capture spatial heterogeneity more effectively.

15.9 Coefficients of Spatial Heterogeneity

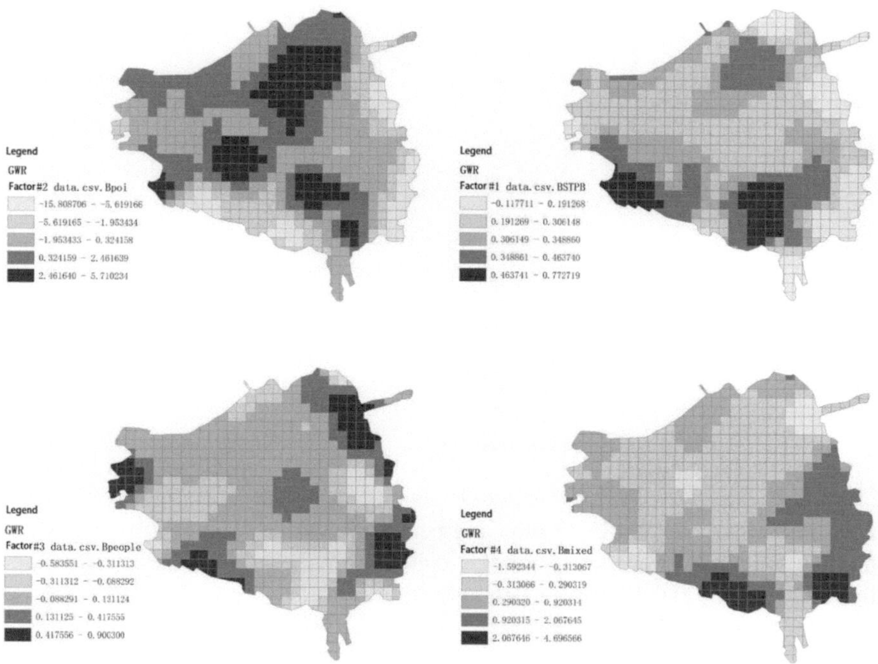

This figure shows the results of geographically weighted regression (GWR) for spatial heterogeneity coefficients across variables like POI density, BSTPB, population density, and land-use mix degree. The color intensity indicates the strength of each variable's impact on shared bike usage, with darker areas reflecting stronger influence and lighter areas showing weaker or negative correlations, highlighting spatial variability across urban regions.

15.10 OD Visualization

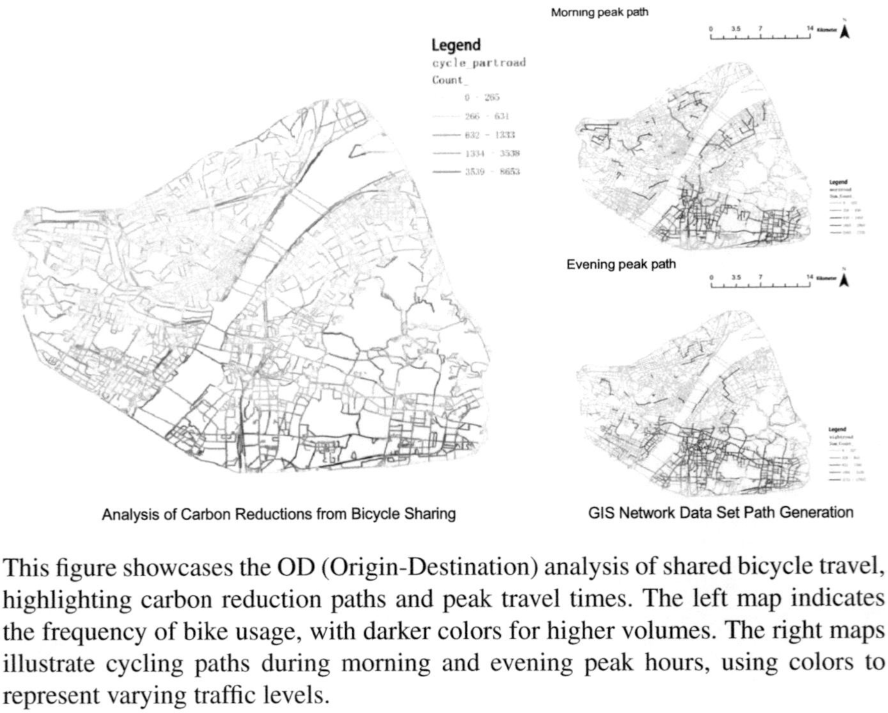

Analysis of Carbon Reductions from Bicycle Sharing　　GIS Network Data Set Path Generation

This figure showcases the OD (Origin-Destination) analysis of shared bicycle travel, highlighting carbon reduction paths and peak travel times. The left map indicates the frequency of bike usage, with darker colors for higher volumes. The right maps illustrate cycling paths during morning and evening peak hours, using colors to represent varying traffic levels.

15.11 Scatters of Morning and Evening Peak Starting Points

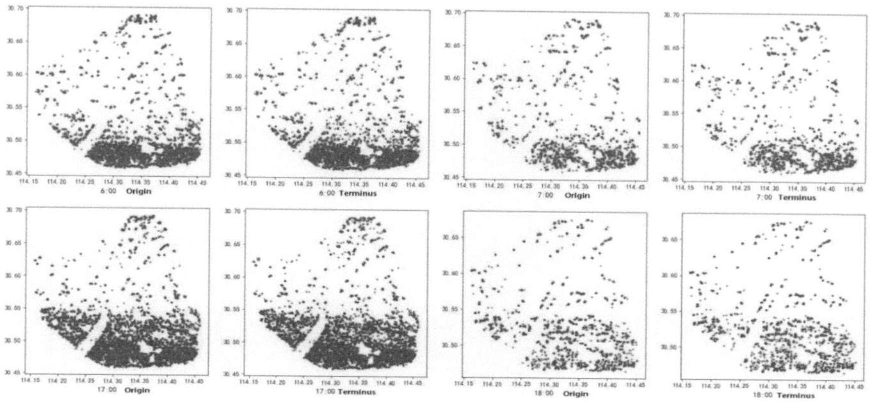

This figure shows the distribution of shared bicycle origin and destination points during morning and evening peak hours. The top four charts represent the morning peak (6:00, 7:00), while the bottom four show the evening peak (17:00, 18:00). Red scatter points indicate trip locations and reveal spatial distribution patterns.

15.12 Projections of Origins and Destinations

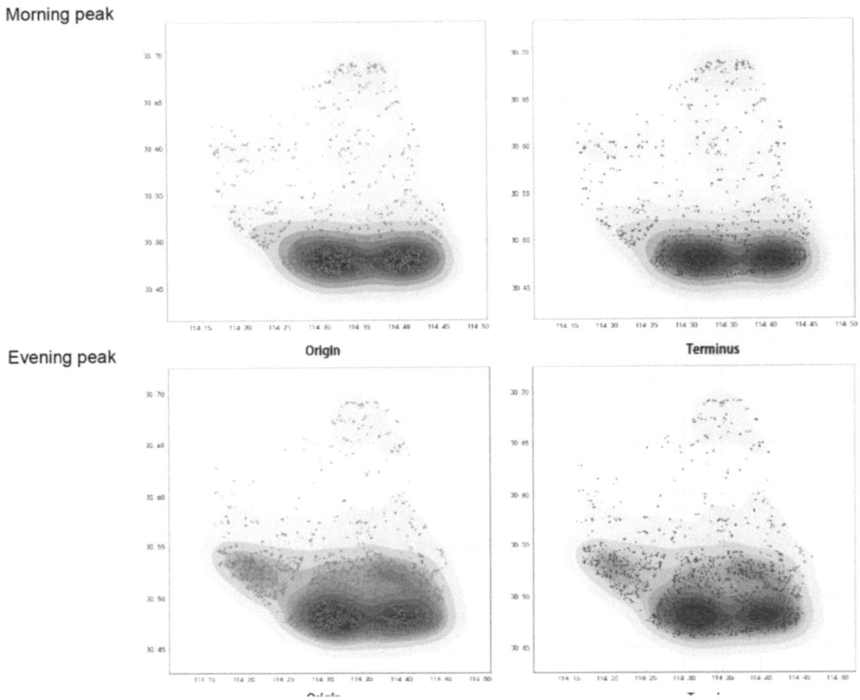

This figure shows the density distribution of shared bicycle origins (Origin) and destinations (Terminus) during peak hours. The top row depicts the morning peak, while the bottom row represents the evening peak. The left charts illustrate origins (red heatmap), and the right charts show destinations (blue heatmap), with darker areas indicating higher travel density.

Part III
DigitalFUTURES 2023: Decoding Morphcarbon Causality

Mentors
Yao Jiawei, Wu Yupeng, Huang Chenyu

Assistants
Anna, Zhang Liwenbo, Wang Jinyu, Wang Yu, Wang Xiaoxin
Workshop Participants: Duan Shufan, Hou Chenfeng, Hu Chenxi, He Mingyi, Ju Chen, Jian Yixin, Li Gaomei, Li Tingting, Liu Yinlu, Ren Zhongyang, Song Junkang, Shen Yanting, Shi Zewi, Tian Hui, Wu Haoran, Ye Jiahong, Yu Mengqing, Zhao Ranting

Topic
In response to the urgent challenges of global climate change, China has pledged to achieve carbon neutrality by 2060. Recognizing cities' significant role in energy consumption and carbon emissions, we've initiated the "Decoding Morphcarbon Causality" workshop to advance this goal. Through interdisciplinary collaboration in architectural sciences, including design, building physics, and HVAC, we'll use AI to analyze the relationship between spatial form, built environment, energy consumption, and emissions. Our aim is to support sustainable urban development.

Content
This workshop will engage in a series of rigorous and cutting-edge educational activities, focusing on fostering interdisciplinary research and innovative thinking among participants, while collectively exploring new methods and approaches to enhance urban sustainability. We encourage bold proposals of innovative viewpoints and concepts, which will be validated and optimized through scientific methods. Through in-depth research and practical exercises, we will gradually uncover the intrinsic connections between spatial forms, built environment, energy consumption, and carbon emissions. Continuously pushing the boundaries of knowledge and uncovering latent potentials, we aim to collectively propel urban architecture towards a sustainable future that is green and low carbon.

Result

This workshop emphasizes the integration of theory and practice, with a strong emphasis on the practical application of research outcomes. In the later stages, we will maintain close cooperation with businesses and government departments to provide participants with platforms for showcasing and applying their research achievements, ensuring that their efforts can truly make a difference in the world! The workshop's results will be presented in the form of posters, short videos, and presentations, allowing participants to fully demonstrate their gains and growth throughout this academic journey.

Review

The workshop focused on fostering interdisciplinary and innovative thinking to enhance urban sustainability. It included cutting-edge educational activities guiding participants to explore new methods and AI algorithms for validation. The aim was to uncover connections between urban built environments, energy consumption, and carbon emissions to support future sustainable city development.

To deepen understanding and encourage cross-disciplinary collaboration, academic salon activities were organized. Experts from various universities gave lectures and thematic presentations. Participants also shared their research findings, enriching the academic atmosphere and promoting collaboration.

The workshop provided theoretical and technical teachings on interpretable machine learning and causal inference, aiming to bridge the knowledge gap in achieving carbon-neutral cities. Mentors and participants engaged in continuous brainstorming, resulting in four outstanding research outcomes.

Trainee Information in 2023

Total applicants: 118
Number of officially admitted trainees: 30
Admission rate: 16.7%

Paper publication

·Ren C, Shi Z, Tian H, et al. Estimating of the Causal Effect of Land Use Mixed on Adult Asthma Prevalence in New York State[J]. Sustainable Cities and Society, 2025: 106125.

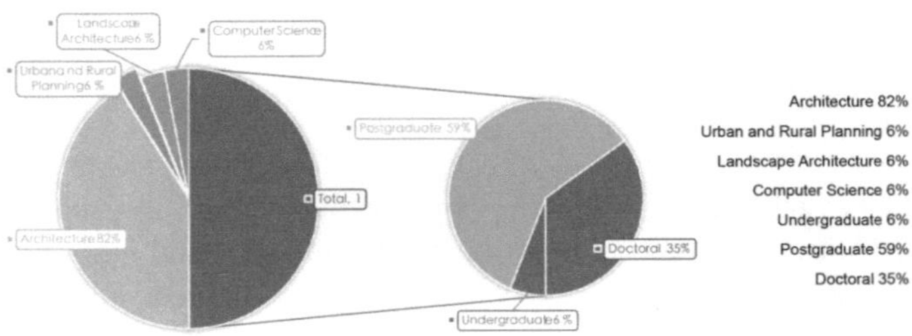

Chapter 16
Carbon Domes—The Impact of Urban Form on Plant Carbon Sinks

Members
Yu Mengqing, Hou Chenfeng, Shen Yanting, Ye Jiahong, He Mingyi

Background
Carbon neutrality is crucial, especially for buildings which are major carbon emitters. Achieving this goal requires action across urban planning, landscaping, energy, transportation, and operations. However, current research often focuses on reducing emissions within specific sectors, neglecting overall carbon balance. To address this gap, we propose a pilot study on carbon-neutral communities, with a focus on urban green spaces as carbon sinks. By starting with Tongji University as a case study, we hope to expand our findings global scope.

Method
1. Campus Model Reconstruction

Using drones to capture footage of Tongji University campus, we collect data for 3D reconstruction. Aerial videos are processed using the Nerf deep learning network system to generate point clouds or mesh surfaces for editing.

2. Calculation of Carbon Sink Disparity

Focusing on the influence of building morphology on plant carbon sequestration, we calculate the disparity between vegetation carbon sink volume in urban structures and biomass-based carbon sink volume. This involves field measurements of biomass on six green spaces at Tongji University and NDVI analysis from satellite images. The campus grid model, reconstructed by drones, is analyzed using ArcGIS, resulting in 290 grid cells. The disparity between actual and biomass-based vegetation carbon sinks is calculated for each cell.

J. Yao, *Intelligent Exploration in Sustainable Design*,
https://doi.org/10.1007/978-981-96-6743-7_16

3. Explainability Analysis

To identify key morphological features affecting carbon sequestration, we input data from 290 groups into a CNN model. We use the CAM explainability analysis method to generate heat maps highlighting which parts of the network contribute most to decision-making during classification tasks.

Significance

Research shows that the spatial distribution of buildings and vegetation significantly impacts plant carbon sequestration. Specifically, in areas where carbon sequestration is positive, dense vegetation and low-rise buildings promote carbon sequestration, likely due to ample sunlight and efficient photosynthesis. Conversely, in areas with negative carbon sequestration, tall buildings and low grassy areas are highlighted, possibly due to building shading and lower plant carbon sequestration capacity. The study suggests that taller buildings restrict vegetation carbon sequestration. Future research needs more detailed spatial indicators to accurately study the impact of urban morphology on plant carbon sequestration.

16.1 Technology Method

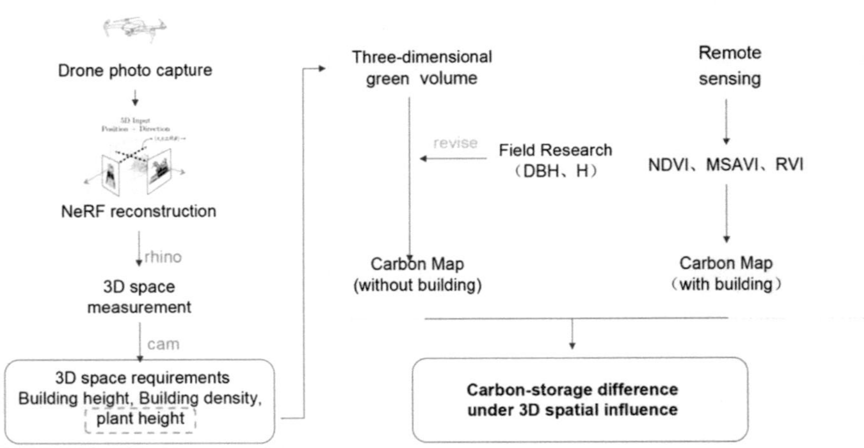

This flowchart outlines the process of calculating urban carbon storage using drone imagery and remote sensing techniques. It begins with drone photo capture, followed by NeRF reconstruction to generate 3D spatial data for measuring building height, density, and vegetation height. The process also incorporates remote sensing methods (NDVI, MSAVI, RVI) to create a carbon storage map that includes buildings. Field research is conducted to calibrate green volume data for a carbon storage map without buildings, allowing for a comparison of carbon storage under different 3D spatial influences.

16.2 Correction of Biomass Data Through Field Measurements

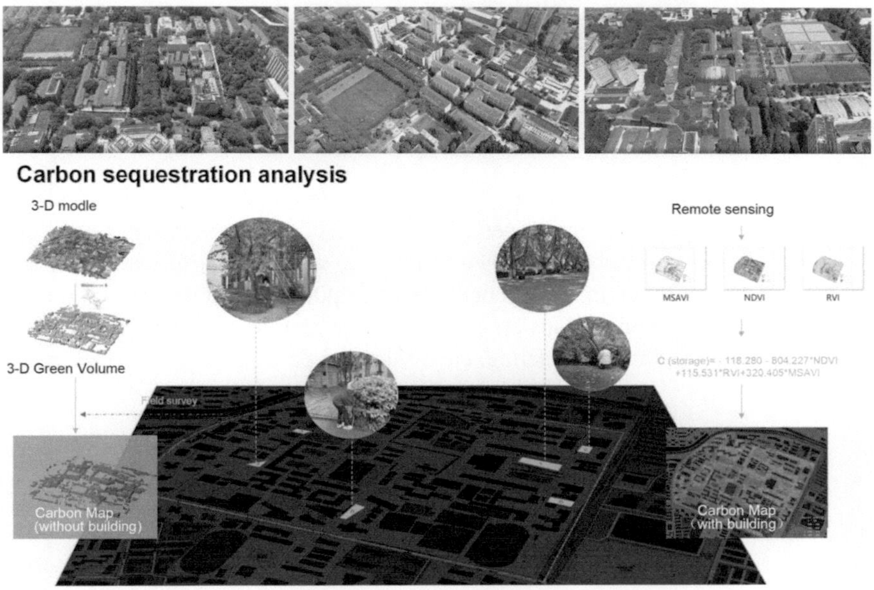

This figure depicts the process of correcting biomass data through field measurements and analyzing carbon sequestration. The top section features drone-captured aerial images used for 3D modeling. The left process corrects data to create a carbon storage map excluding buildings, while the right integrates remote sensing data (MSAVI, NDVI, RVI) to include buildings in the carbon storage map. These datasets are then used to evaluate carbon sequestration capacity across various urban spatial structures.

16.3 Nerf 3D Reconstruction Technology

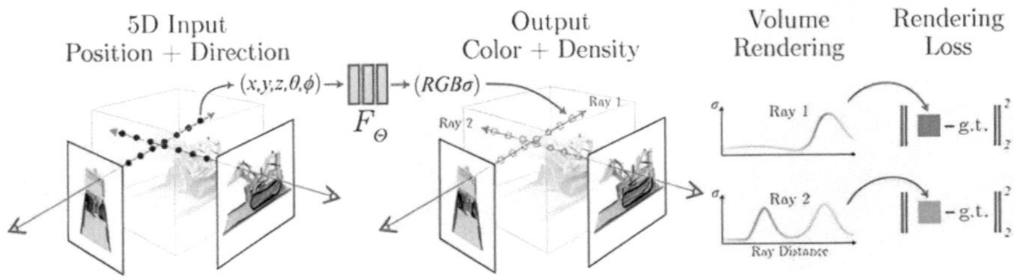

This figure illustrates the basic process of NeRF (Neural Radiance Fields) 3D reconstruction technology. NeRF uses a neural network to implicitly represent a 3D scene by taking 5D input information (position (x, y, z) and viewing direction (θ, φ)), which is mapped through the neural network to output color (RGB) and volumetric density (σ). This method enables efficient reconstruction of complex scenes and is widely applied in 3D urban modeling, environmental perception, and virtual reality applications.

16.4 Campus Digital Model

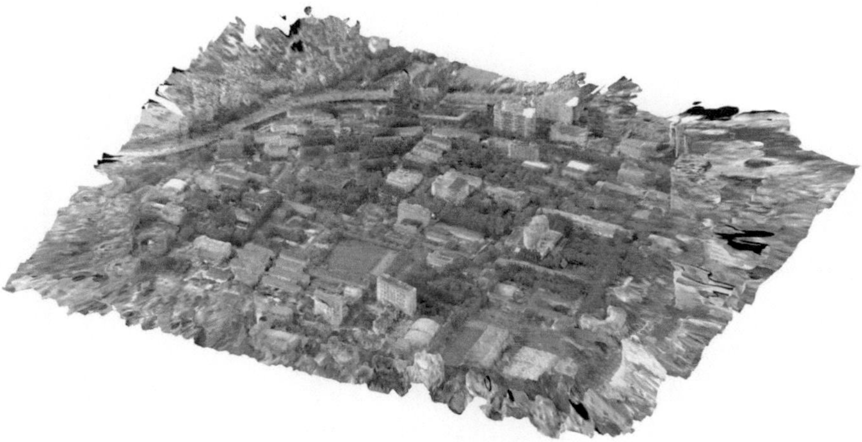

3D reconstruction model of Tongji University

A 3D reconstruction model of Tongji University, generated using drone imagery and NeRF technology, accurately restoring the campus terrain, buildings, and green spaces.

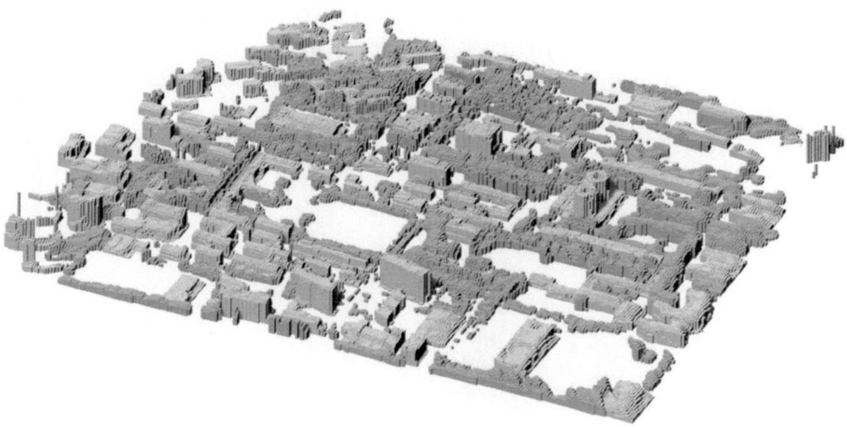

Building Vegetation Segmentation Model

A building-vegetation segmentation model, where buildings (gray) and vegetation (green) are classified separately to analyze the impact of different spatial elements on carbon storage. This model can be applied to urban green space management, carbon sink calculations, and spatial planning optimization.

16.5 Machine Learning Models for Predicting Carbon Differences

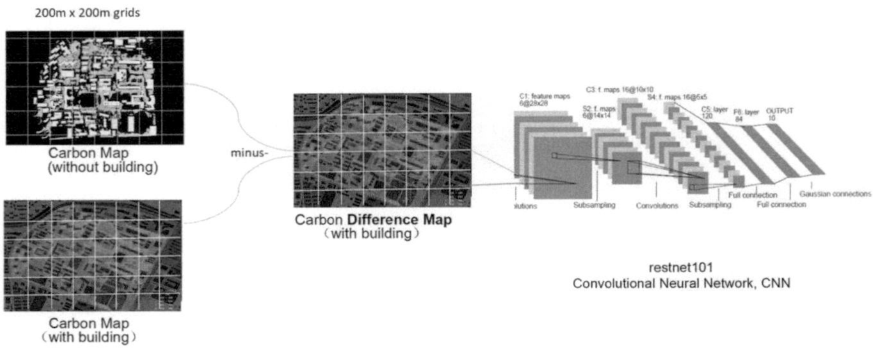

Machine learning model: By comparing carbon storage maps without buildings and with buildings, the Carbon Difference Map is calculated. The ResNet101 Convolutional Neural Network (CNN) is used for training to extract features and predict changes in carbon storage.

16.6 Explainable Analysis

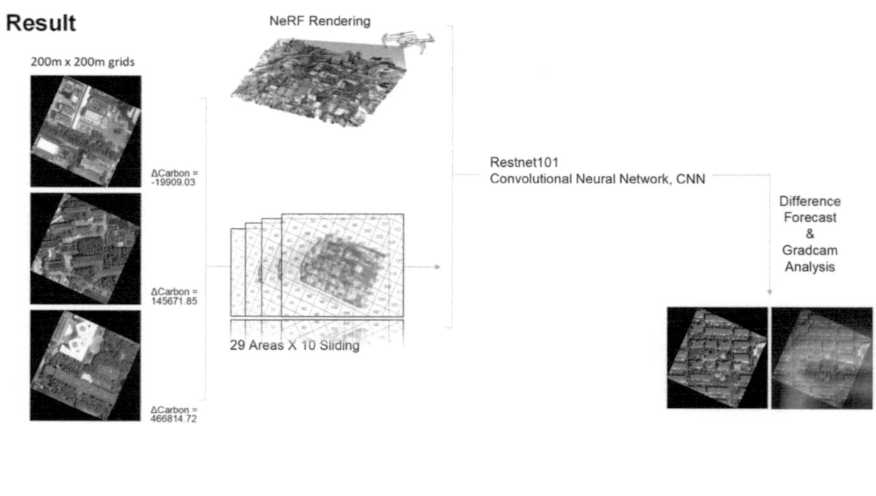

Explainable Analysis: NeRF rendering generates high-precision 3D scenes, with urban areas divided into 200 m × 200 m grids to calculate carbon storage changes (ΔCarbon). Grad-CAM analysis visualizes model focus areas and interprets spatial patterns of carbon storage differences.

16.7 Result

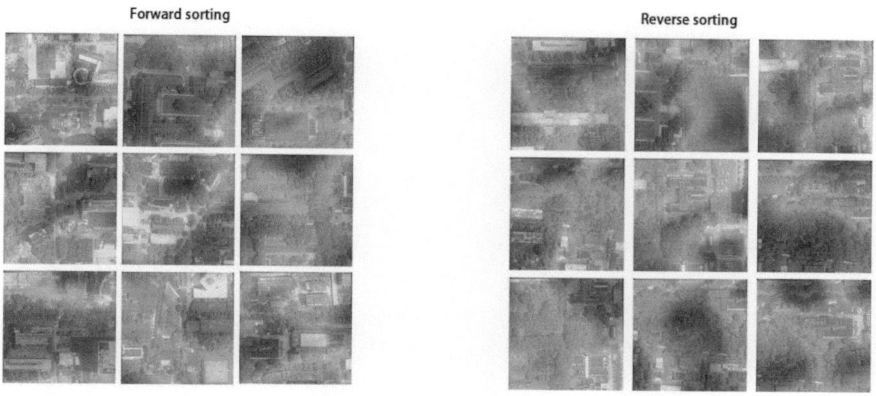

This figure presents the visualization results based on Grad-CAM analysis, ranking regions by carbon storage variation.

Left (Forward sorting): Sorted in ascending order of carbon storage variation, where red areas indicate higher carbon storage and blue areas indicate lower carbon storage.

Right (Reverse sorting): Sorted in descending order, highlighting the distribution of low carbon storage areas.

Chapter 17
Changing with the Weather: Climate-Interactive Facade Reduction of Carbon Emissions

Members
Wu Haoran, Wang Xiaoxin, Li Tingting, Jian Yixin

Background
Climate-Interactive Facade refers to an architectural envelope that adjusts its form and function in real-time to optimize multiple environmental performance metrics. A well-designed facade can reduce building energy consumption and carbon emissions by 20–30%. This study focuses on dynamic sunshades and aims to investigate their impact on carbon emissions in Shanghai office buildings when deployed on a large scale.

Method
1. Dynamic Facade

Create a triangular dynamic facade by obtaining normalized hourly values from window solar irradiance, indoor average illuminance, and outdoor dry-bulb temperature. Calculate predetermined functions to derive five distinct dynamic shading states.

2. Optimization Objectives

Simulate the dynamic facade model to obtain metrics like Useful Daylight Illuminance (UDI), Energy Use Intensity (EUI), Thermal Discomfort Probability (TDP), and photovoltaic power generation. Use these metrics as optimization objectives for multi-objective optimization, considering daylighting, energy consumption, and thermal comfort. Obtain various dynamic shading schedules and annual carbon emission reductions per building unit based on optimal solutions.

3. Calculation for Shanghai Offices

Estimate annual carbon emission reductions per building unit from individual dynamic shading schedules. Use Baidu Maps to collect Shanghai city maps and

J. Yao, *Intelligent Exploration in Sustainable Design*,
https://doi.org/10.1007/978-981-96-6743-7_17

building type ranges, focusing on office buildings. Calculate the total annual carbon emission reduction for Shanghai's office buildings by determining the effective facade area and annual carbon emission reductions per unit area.

4. Real-time Prediction of Shading Schedules

Train a Deep Neural Network (DNN) model with different dynamic shading schedules and weather forecast data. Use the trained model to generate hourly-scale real-time predictions of future dynamic shading schedules, facilitating year-round operational planning.

Significance
After simulation, optimization, and prediction, the optimal operation of dynamic shading in Shanghai office buildings results in an annual carbon reduction of about 21 kg per square meter of unobstructed south facade. Considering Shanghai's total effective facade area of 3,304,900 square meters, the total annual carbon reduction for office buildings is estimated to be 31,231.305 metric tons.

17.1 Dynamic Facade

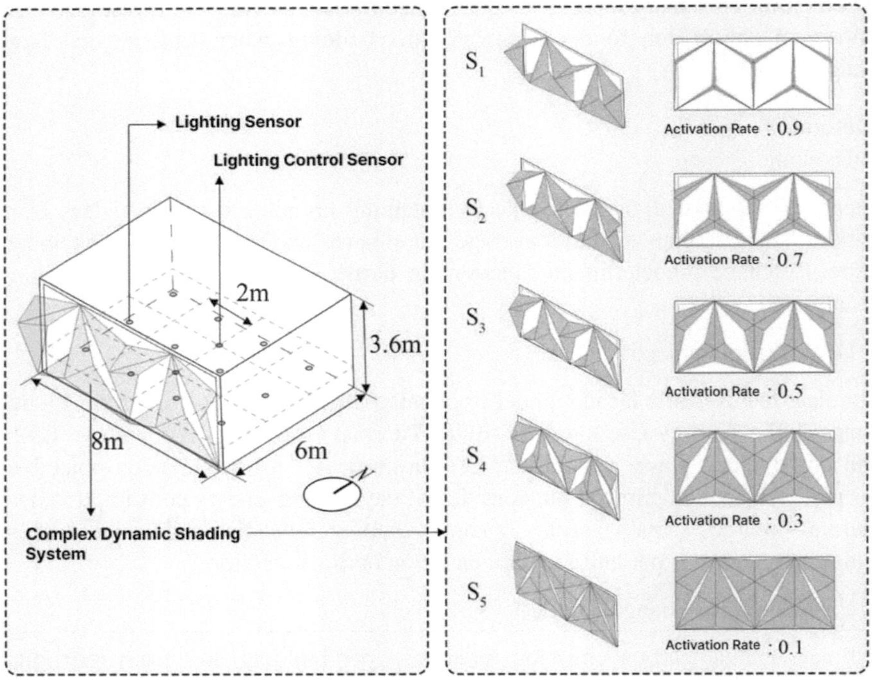

This figure shows the optimization of a dynamic facade, using a complex shading system to adjust exterior shading. Left: The building model has lighting sensors and control sensors to monitor light and adjust shading in real time.

Right: Shading activation rates (0.9 = max, 0.1 = min) adapt to sunlight and energy goals.

The system intelligently regulates shading, improving energy efficiency, indoor lighting, and reducing energy use.

17.2 Research Scope

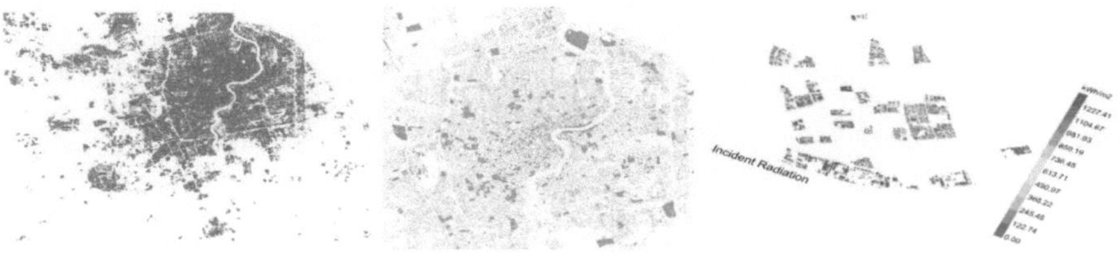

This study analyzes Shanghai's urban blocks, evaluating building/street layouts and solar radiation distribution. Left: city-scale study scope. Middle: detailed building structures. Right: heat map of solar radiation intensity on building surfaces. Supports dynamic facade optimization, improving energy efficiency and indoor comfort.

17.3 Technology Method

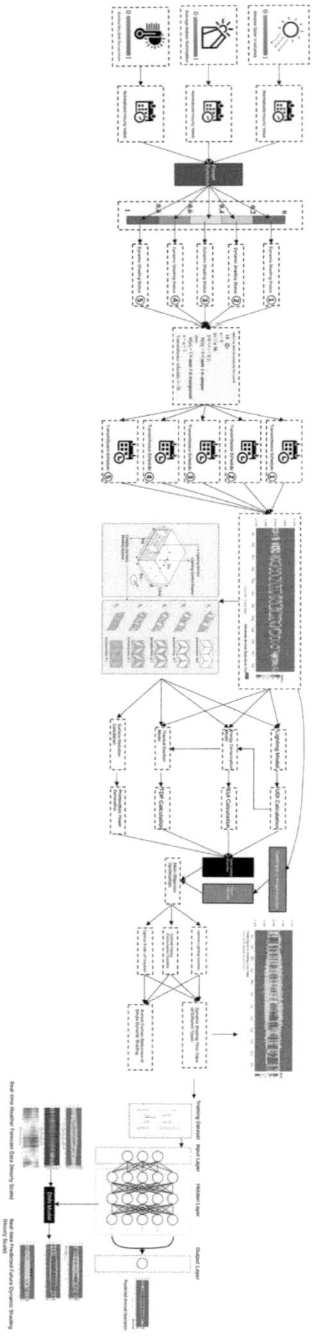

This study analyzes Shanghai's urban blocks, evaluating building/street layouts and solar radiation distribution. Left: city-scale study scope. Middle: detailed building structures. Right: heat map of solar radiation intensity on building surfaces. Supports dynamic facade optimization, improving energy efficiency and indoor comfort.

17.4 Cause-Effect Inference

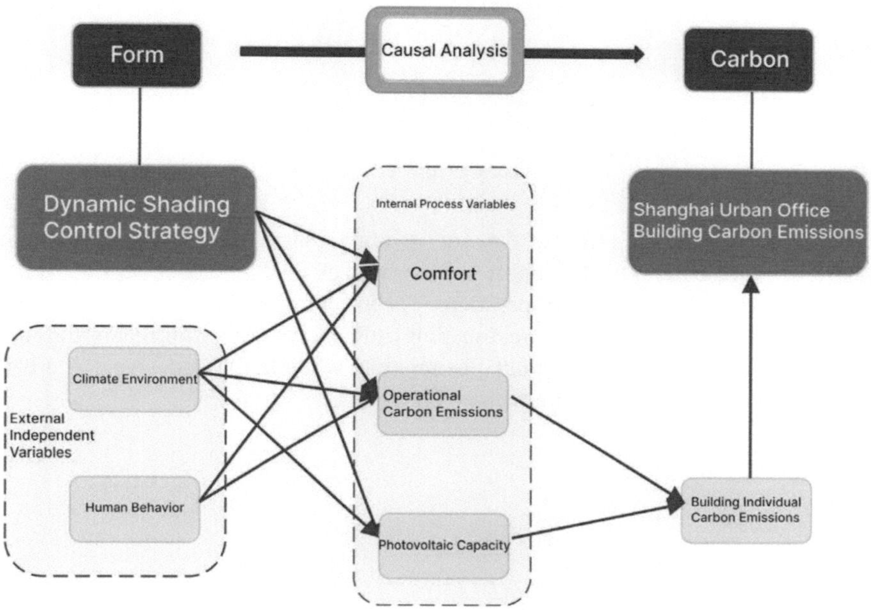

This figure analyzes how dynamic shading control strategies impact carbon emissions in Shanghai office buildings. External factors like climate and human behavior influence shading adjustments, which in turn affect comfort, operational carbon emissions, and photovoltaic capacity, ultimately reducing overall carbon emissions and aiding in the optimization of shading strategies.

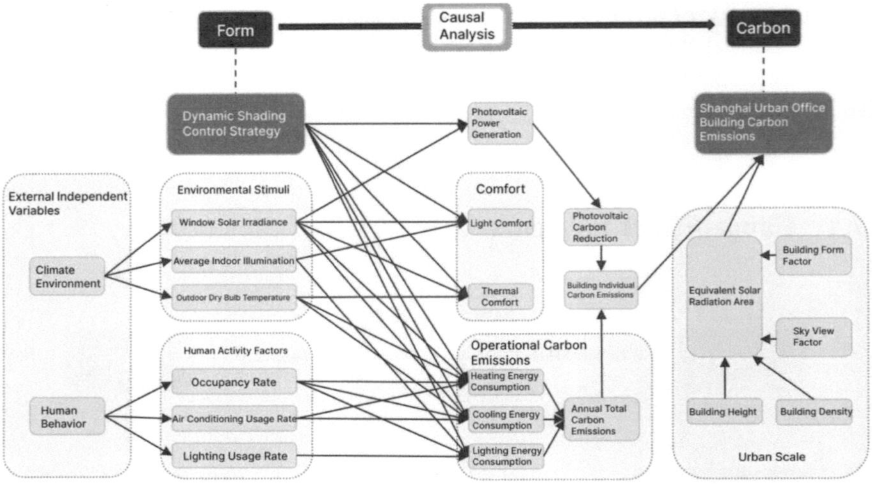

This figure analyzes the effect of dynamic shading control on carbon emissions in Shanghai office buildings. The upper section utilizes a DNN model with real-time weather data to enhance shading operations. The lower section highlights how climate conditions and human behavior affect shading, influencing carbon emissions through comfort, energy use, and photovoltaic generation, while building design and urban solar radiation ultimately determine total emissions.

Chapter 18
CausalAI Reveals Mechanisms Driving Healthiness in Urban Transportation Spaces

Members
Tian Hui, Ren Zhongyang, Zhao Ranting, Shi Zewi, Hu Chenxi

Background
Previous studies have shown that vehicular emissions are the primary source of air pollution and urban CO_2 emissions in non-industrialized cities, strongly linked to respiratory and chronic diseases like asthma and COPD. We ask: Does urban road layout impact respiratory and chronic disease risks? How are socio-economic and environmental factors related to road layout? How can AI help adjust road parameters to reduce respiratory disease risks for nearby residents?

Method
1. Data Collection

Gather risk maps of boat morbidity and chronic obstructive pulmonary disease (COPD) for the Boston area in 2018 (based on census data). Collect vector data of the urban road network in the Boston area and perform clustering of urban areas using descriptive variables such as Road_IN_Bumber, E_charger number, L_LENGTH, L_COUNT, N_LENGTH, N_COUNT, R_LENGTH, R_COUNT, US_LENGTH, US_COUNT, SR_LENGTH, and SR_COUNT to facilitate subsequent discussions.

J. Yao, *Intelligent Exploration in Sustainable Design*,
https://doi.org/10.1007/978-981-96-6743-7_18

2. Cluster Analysis

Use 12 road network morphology variables for Kmeans unsupervised clustering calculations, determining the optimal number of clusters as three using the "elbow method." Establish machine learning models including Support Vector Machine (SVM), Gradient Boosting Machine (GBM), Decision Tree (DT), Random Forest (RF), K-Nearest Neighbor (KNN), and Light Gradient Boosting Machine (LGBM) for each cluster. The second cluster shows the best performance across all models.

3. Interpretability Analysis

Apply the SHAP (SHapley Additive exPlanations) method for interpretability analysis. Generate SHAP importance analysis plots from the best machine learning models to identify nine important correlated factors. Project their importance onto GIS geographical images to obtain geographic weights for these factors.

Significance

After adjusting for socio-economic factors, regression models linking road layout participation to respiratory diseases show notably lower fitness compared to unfitted models. However, road area remains the most influential parameter. Road area contributes to 20%–30% of the most relevant interval for both COPD and asthma diseases. The combination of road area with secondary or tertiary road lengths is the most critical factor affecting disease peaks or underestimations. Features like bus stops or charging stations correlate with reduced disease risk, but the combined impact of parameters on the model is yet to be explored.

18.1 Technology Method

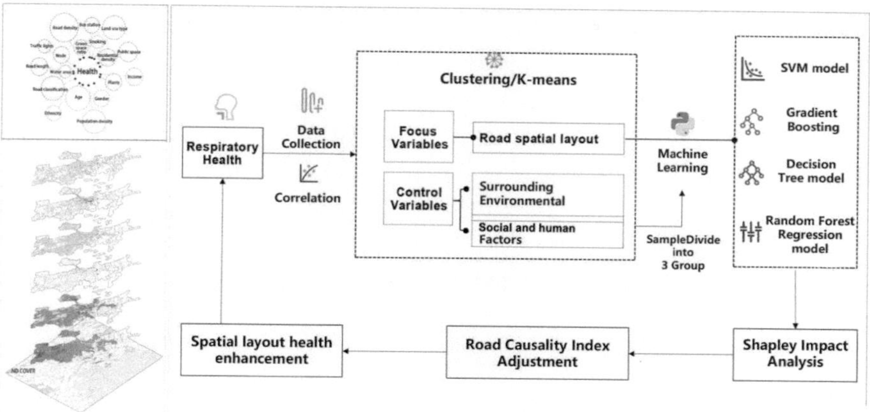

This study examines the link between street layout and respiratory health. It gathers data to correlate road designs with environmental and social factors affecting health. K-means clustering categorizes these variables, and machine learning models (SVM, Gradient Boosting, Decision Tree, and Random Forest Regression) are utilized for causal analysis. Shapley impact analysis adjusts the road causality index to enhance urban planning and public health.

18.2 Research Target

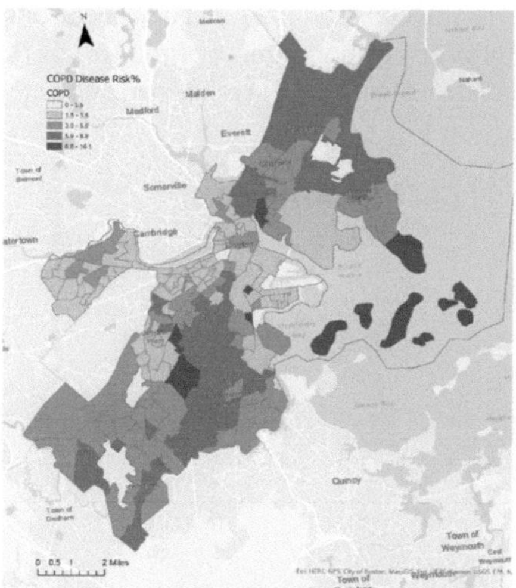

2018 Boston asthma prevalence risk map

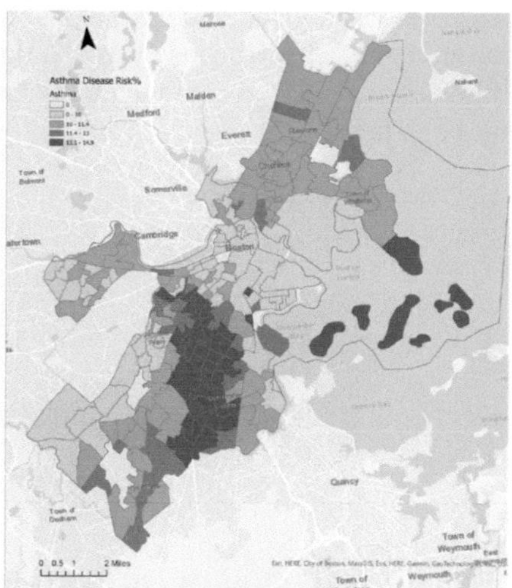

2018 Boston COPD prevalence risk map

This study selects Boston as the research site to analyze the spatial distribution of asthma and chronic obstructive pulmonary disease (COPD) in 2018. The left map illustrates asthma prevalence risk, while the right map shows COPD prevalence risk, with color intensity representing varying disease risks. This data is used to explore the relationship between street layout and respiratory health.

18.3 Causal Inference

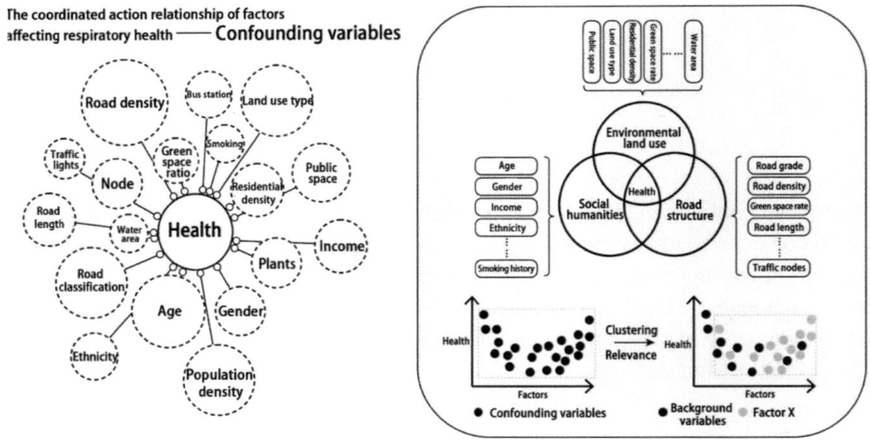

This figure examines the relationship between road structure, environmental land use, and social factors on respiratory health. It highlights confounding variables such as road density and population demographics that affect health. Influencing factors are categorized into three areas: environmental land use, social aspects, and road structure. Clustering analysis identifies the correlations among these variables, differentiating between confounding and key factors to assess the road environment's impact on health.

18.4 Correlation Matrix

This figure is a correlation matrix, showing the relationships between road characteristics, land use, demographic data, and health indicators (such as COPD and asthma). The color represents correlation strength, with green indicating positive correlation and blue indicating negative correlation, while the numerical values represent Pearson correlation coefficients, helping to identify key factors that may impact respiratory health.

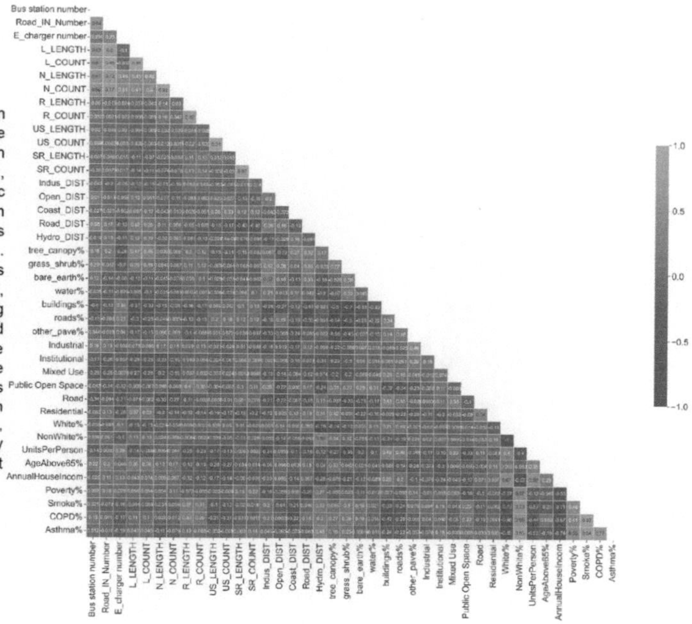

18.5 K-Means Clustering

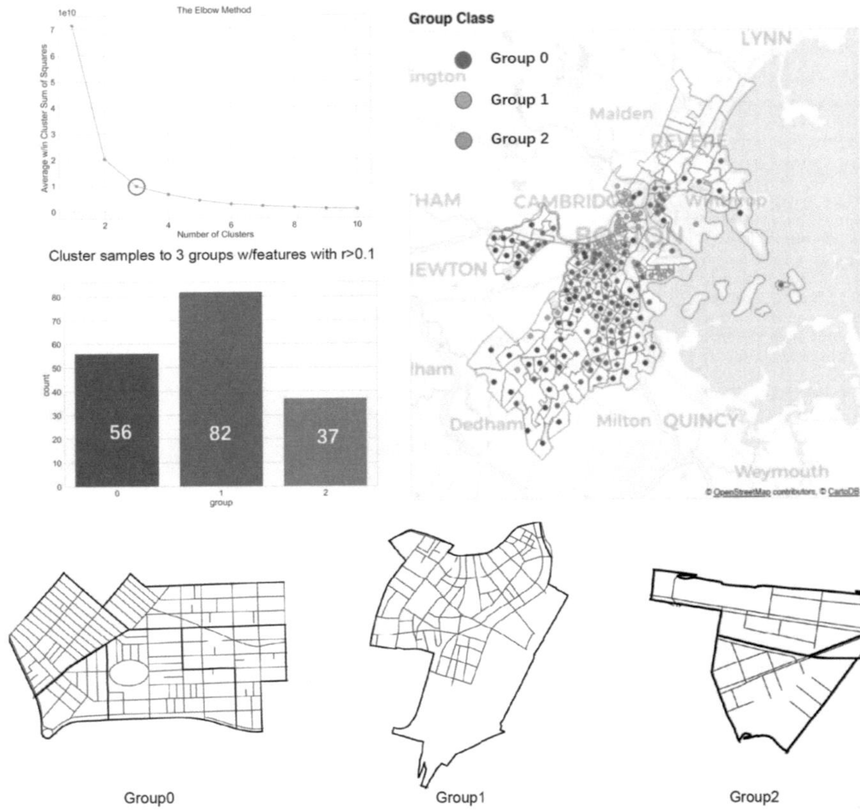

This figure presents K-Means clustering analysis, dividing the study area into three groups based on feature correlation ($r > 0.1$). The Elbow Method (top left) determines the optimal number of clusters, while the bar chart shows the sample count for each group. The map on the right illustrates the spatial distribution of different clusters, and the bottom diagrams compare the street structure characteristics of each group to explore the potential impact of different street patterns on respiratory health.

18.6 Machine Learning

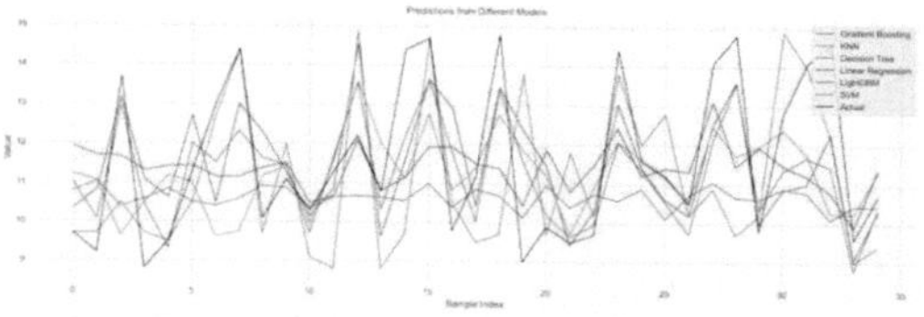

Asthma - Machine Learning Supervised Learning Regression Comparison

		LGBM	GMB	DT	RF	SVM	KNN
G1	R2	-0.03	0.17	-0.37	0.18	-0.09	0.03
	RMSE	1.04	0.93	1.20	0.93	1.76	1.05
G2	R2	0.02	0.19	-0.26	0.24	0.16	0.12
	RMSE	1.62	147	1.85	1.43	1.50	1.54
G3	R2	/	/	/	/	/	/
	RMSE	/	/	/	/	/	/

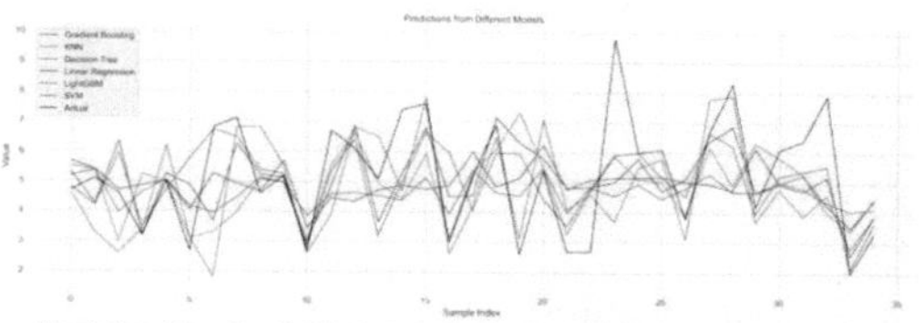

Chronic Non-obstructive - Machine Learning Supervised Learning Regression Comparison

		LGBM	GMB	DT	RF	SVM	KNN
G1	R2	0.20	0.18	-0.37	0.11	-0.02	0.05
	RMSE	1.75	0.12	1.20	1.05	1.12	1.05
G2	R2	0.00	0.35	-0.13	0.23	0.0	0.22
	RMSE	2.21	1.70	2.26	1.86	2.22	1.83
G3	R2	-0.0	0.21	-0.64	0.11	0.43	0.0
	RMSE	0.59	0.67	0.97	0.75	0.57	0.72

This figure presents a comparison of machine learning models for predicting asthma and chronic non-obstructive disease regression.

The line charts on the left compare the prediction performance of different models (LGBM, GMB, DT, RF, SVM, KNN), while the tables on the right list the R^2 values and RMSE errors for each model across different groups (G1, G2, G3). The results indicate variations in prediction accuracy across different diseases and clusters, providing insights into the impact of different street patterns on respiratory health.

18.7 Contribution Analysis of Shap Model of Asthma

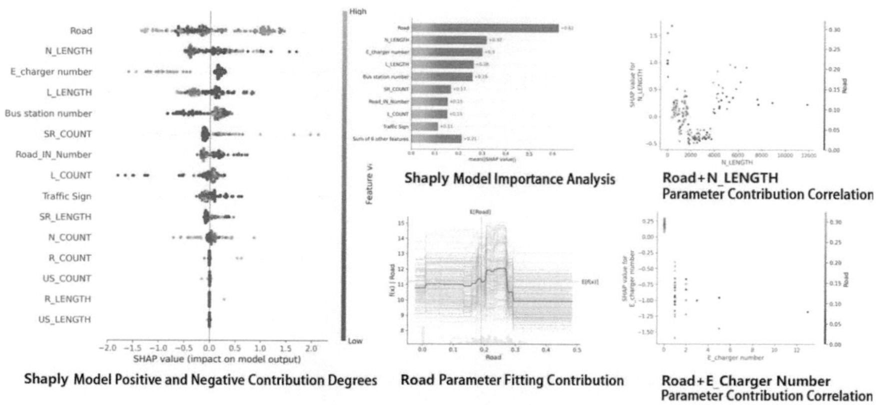

This figure presents a SHAP analysis of asthma risk related to road variables. The left plot shows the positive and negative contributions of these variables, while the center quantifies their influence. The right scatter plot illustrates the correlation between key variables (e.g., N_LENGTH, E_charger number) and their contributions. The fitted curve at the bottom visualizes how road parameters impact asthma risk.

18.8 Contribution Analysis of SHAP Model of COPD

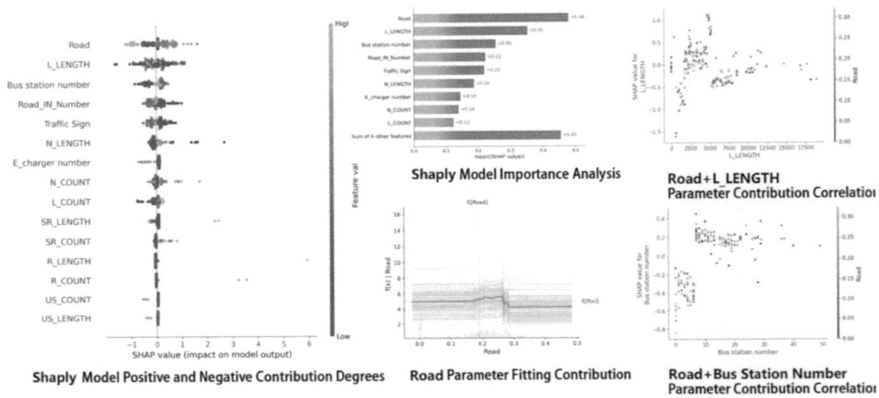

This figure shows the SHAP model analysis for COPD, highlighting how road-related variables affect COPD risk. The left side displays positive and negative contributions, the center quantifies variable importance, and the right side visualizes correlations between key variables and model contributions, illustrating the impact of road environments on respiratory health.

18.9 Important Correlation Factors Geographical Weighting

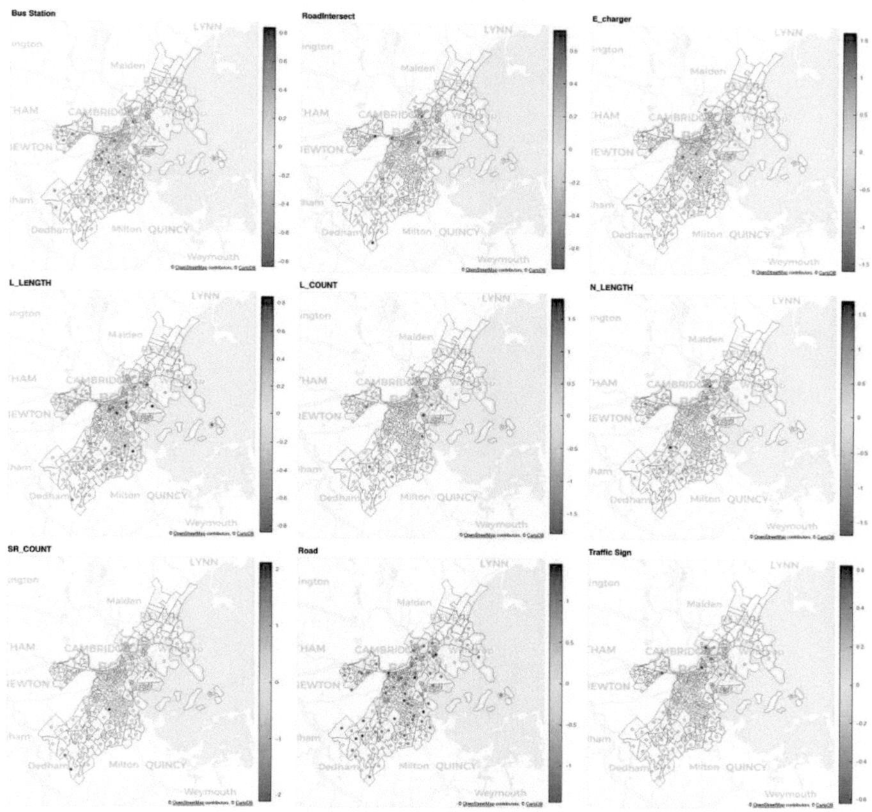

This figure shows the spatial visualization of SHAP analysis results, illustrating the geographic impact of road-related variables on COPD risk. Colors indicate SHAP value distribution, with red for positive contributions and blue for negative. The variations reveal how road environmental factors affect respiratory health regionally.

Chapter 19
Optimization of Space-Energy Coupling in Future Climate-Connected City Neighborhoods

Members
Li Gaomei, Ju Chen, Song Junkang, Liu Yinlu, Duan Shufan

Background
Over the past decade, global temperatures have surged by 1.09 °C compared to the late nineteenth century, marking unprecedented levels. The last five years have been record-breaking in terms of heat. Scientists predict further increases of over 1.5 °C by the mid-2030s. While factors like urban climate, human behavior, building equipment, and neighborhood design all affect energy use, the specific impact of neighborhood layout remains unclear. This study aims to optimize urban neighborhood layouts for energy efficiency, maximizing photovoltaic potential and minimizing carbon emissions. It explores how different weather conditions affect optimal designs to find solutions with the lowest energy use and emissions.

Method
1. Construct Urban Dataset

Collect future weather files (2020–2080) and extract prototype neighborhoods. Set morphological parameters and create a parametric model for neighborhood generation using Latin Hypercube Sampling.

2. Performance Simulation

Use Rhino's Grasshopper platform with Ladybug and Dragonfly plugins to simulate solar potential, energy consumption, and carbon emissions of parametric neighborhood models.

3. Machine Learning Algorithms

Train BP neural networks and ensemble learning algorithms with simulated target data and neighborhood parameters to predict sunlight irradiation and building energy

J. Yao, *Intelligent Exploration in Sustainable Design*,
https://doi.org/10.1007/978-981-96-6743-7_19

consumption. Utilize a multi-objective genetic algorithm in the WallaceiX platform to optimize neighborhood generation considering carbon emissions, solar radiation, and cooling ratios under different weather conditions.

4. Explainable Analysis.

Conduct SHAP-based interpretability analysis on morphological parameters and carbon emission datasets. Generate ICE and PDP analysis plots to understand how indicators influence neighborhood performance.

Significance

The research finds that mixed-use neighborhoods should prioritize residential areas over office and commercial spaces. With future climate changes, cooling energy demand rises while heating energy decreases. While buildings and solar potential partly meet energy needs, they can't fully cover them. A 7:1:1 ratio for residential, office, and commercial functions works best. Neighborhood layouts should adjust to changing climates, emphasizing lifecycle performance optimization in planning and design.

19.1 Technology Method

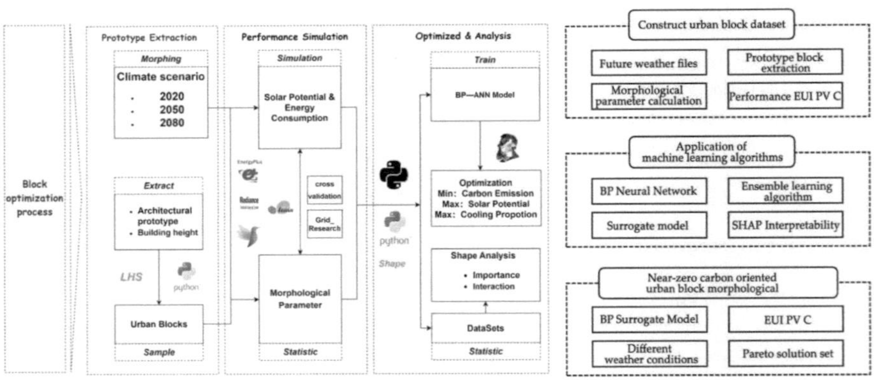

This diagram outlines a method to optimize urban block morphology for future climate scenarios (2020, 2050, 2080) through morphological extraction, performance simulation, and optimization analysis. It involves extracting architectural prototypes and heights, simulating solar potential and energy consumption, and using the BP-ANN neural network for optimization to reduce carbon emissions and enhance solar potential and cooling. Lastly, SHAP interpretability analysis assesses the significance and interactions of morphological parameters to support near-zero carbon urban blocks.

19.2 Research Conception

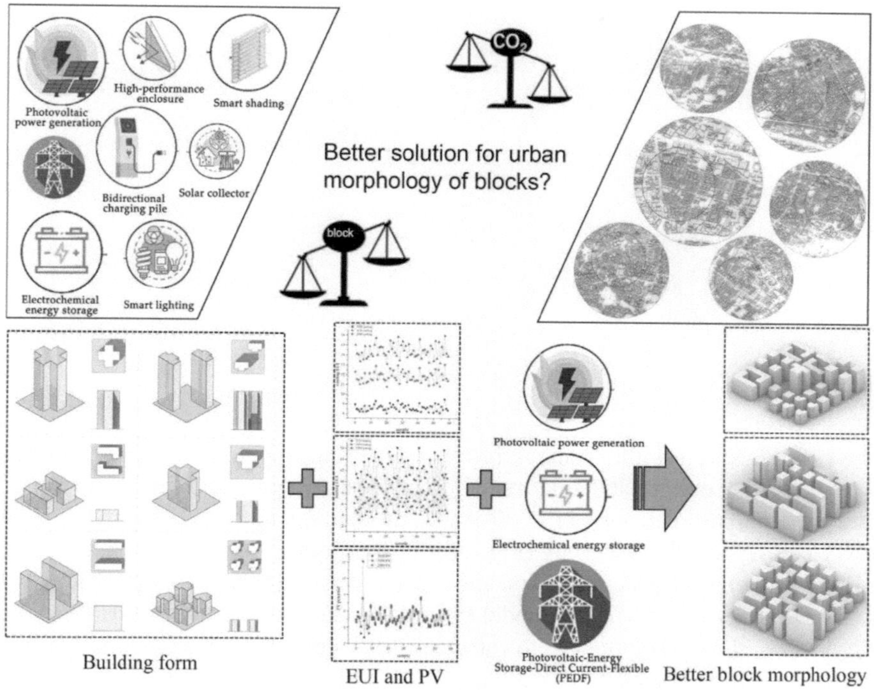

This diagram outlines a research concept for optimizing urban block morphology to boost energy efficiency and cut carbon emissions. It examines variations in building form, energy use intensity (EUI), photovoltaic power generation (PV), and advanced solutions like energy storage systems and smart shading. The aim is to balance urban design with CO_2 emissions, identifying sustainable urban block morphologies that maximize renewable energy use.

19.3 Research Process

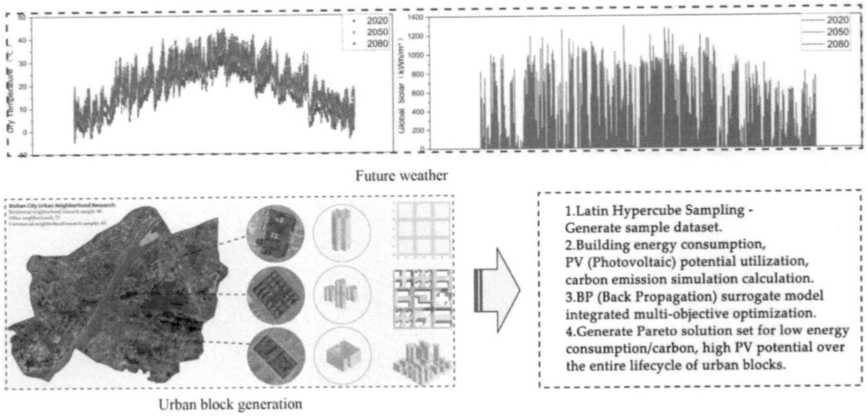

The diagram outlines the research process, featuring future climate data analysis (including temperature and solar radiation changes), urban block generation from various neighborhood samples in Wuhan, and optimization modeling. The study uses Latin Hypercube Sampling (LHS) for data generation, integrates photovoltaic utilization with energy consumption and carbon emission simulations, and employs a BP neural network surrogate model for multi-objective optimization. Ultimately, it produces a Pareto solution set to enhance energy consumption and reduce carbon emissions in urban blocks.

19.4 Result and Discussion

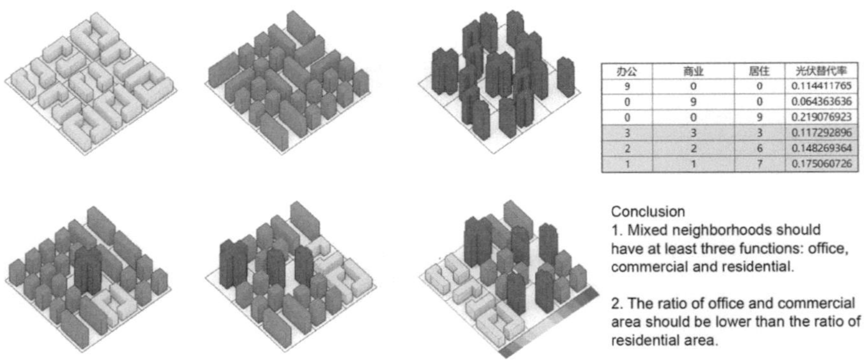

The figure showcases various urban block layout schemes with different functional combinations and their effects on photovoltaic substitution rates. It outlines six schemes, including single-function blocks (residential, office, commercial) and mixed-function blocks with varied proportions of these spaces.

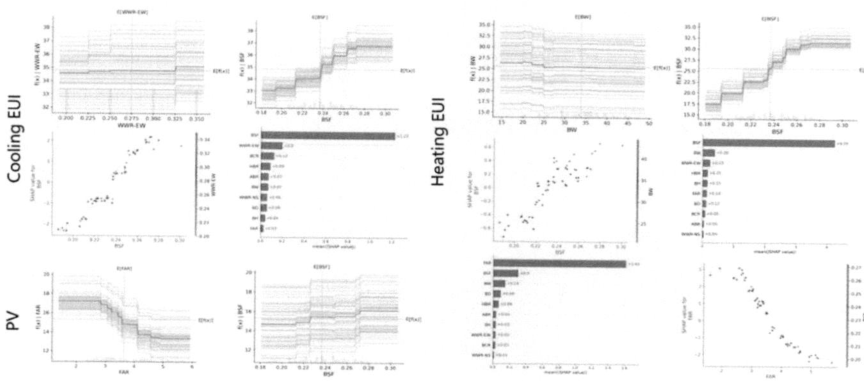

1. The most affected morphological parameters for cooling EUI, heating EUI and PV were BSF and WWR-EW; BSF and BW; FAR and BSF, respectively.
2. Non-linear relationship between morphological parameters and Cooling EUI, Heating EUI and PV.

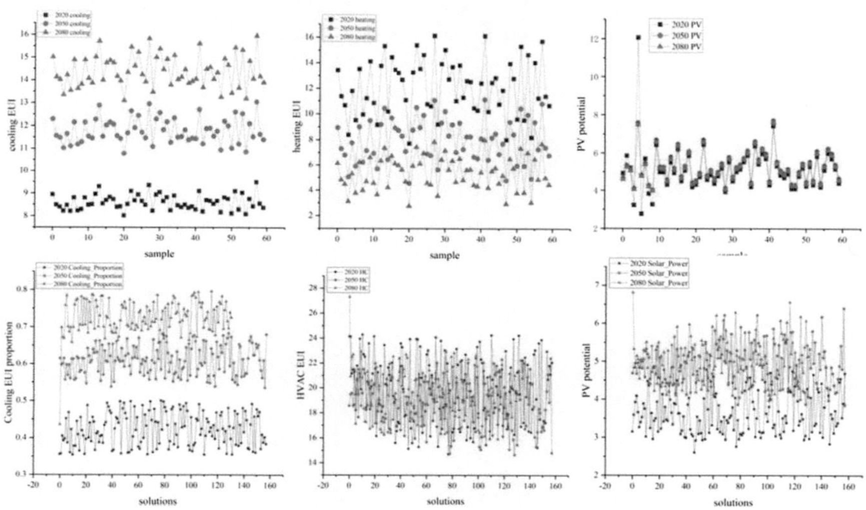

1. Cooling EUI increases with weather, heating EUI decreases.
2. PV potential increases with time.

This figure shows how morphological parameters impact cooling EUI, heating EUI, and photovoltaic (PV) potential. Cooling EUI is most influenced by BSF and WWR-EW, heating EUI by BSF and BW, and PV potential by FAR and BSF. The study also highlights a nonlinear relationship among these parameters. As the climate changes, cooling EUI increases, heating EUI decreases, and PV potential improves.

19.5 Multi-objects Optimization

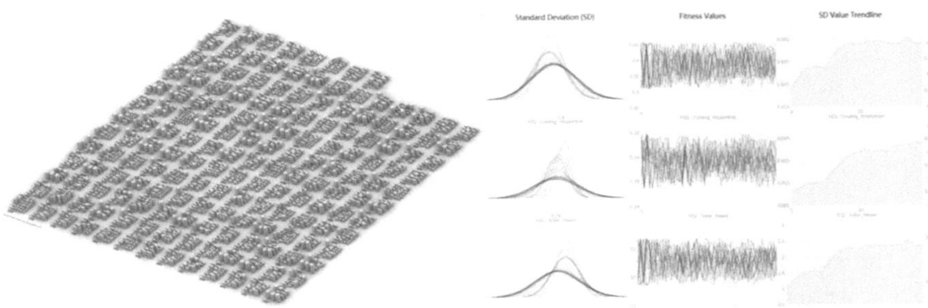

This figure illustrates multi-objective optimization for urban block morphology. The left side shows parametric urban form variations, while the right side analyzes optimization performance using statistics. The process balances cooling proportion, solar power use, and other factors to identify sustainable design solutions.

Part IV
DigitalFUTURES 2024: Mentors

Mentors
Yao Jiawei, Sergio Altomonte Maider, Llaguno-Munitxa, Huang Chenyu

Assistants
Wang Jinyu, Zhong Zhengjia, Ye Jiahong, Liu Yu, Wang Chenyu

Workshop Participants
Du Shaozi, Yan Hongming, Xu Yuhan, Qin Yuchen, Xie Yuchen, Chen Kaifan, Hou Chenfeng, Zhang Yuling, Ning Qingru, Lin Dianyi, Xu Nuoyan, Wei Xinran, Deng Zixuan, Meng Mei, Lei Zhen

Background
With the continuous advancement of urbanization, sustainable development has become a global focus. The multiphysics problems in building physics and urban environments, such as fluid dynamics, heat transfer, mass transfer, and the urban heat island effect, are particularly complex and critical. This workshop aims to explore and implement methods for the reconstruction of multiphysics fields using artificial intelligence algorithms to support scientific discovery and engineering optimization

Content
Participants will engage in a cutting-edge artificial intelligence algorithm competition, where they will learn about the key challenges in the field of building science and explore ways to apply artificial intelligence. The competition will be divided into two tracks: "Scientific Discovery" and "Scientific Simulation," aiming to explore methods for reconstructing multiphysics fields in the intelligent era.

(1) Scientific Discovery Track: Focuses on energy and environmental issues in sustainable cities, uncovering hidden relationships between variables.
(2) Scientific Simulation Track: Focuses on performance simulation in green and low-carbon building design, creating efficient and robust surrogate models.

Each track will release dedicated datasets collected from real scientific research scenarios. Participating teams will need to independently develop algorithms to address the challenges of reconstructing multiphysics fields in urban buildings. An online platform will be provided, where teams can submit algorithm code and preliminary results, receive real-time feedback, and evaluate performance. Evaluation criteria will include: technical innovation, solution quality, scalability, and practicality.

Forum Topics
This workshop provides participants with a wealth of theoretical and technical educational resources, ensuring that even those with no prior background can quickly build problem-solving algorithms through guidance. The workshop offers an inclusive collaborative platform, encouraging researchers and engineers from diverse disciplinary backgrounds to collectively tackle the forefront challenges of multiphysics field reconstruction. We are organizing several high-quality forums, including invited lectures, a young scholars' forum, and a youth sharing session. This workshop aims to break down disciplinary barriers and promote the exchange of ideas.

Trainee Information in 2024

Total applicants: 102

Number of officially admitted trainees: 15

Admission rate: 14.7%

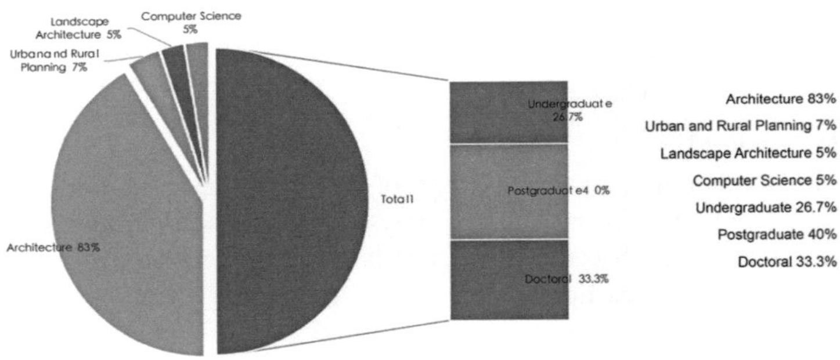

Chapter 20
Multi-objective Optimization of Environmental Equity and Synergistic Benefits of "Carbon-Pollution-Heat" at the Urban Plot Scale: Based on Multimodal Deep Learning

Members
Yan Hongming, Meng Mei, Wei Xinran, Du Shaozi, Xu Yuhan

Background

Carbon-pollution synergy refers to the simultaneous reduction of carbon emissions (mainly CO2) and air pollutants (like PM2.5, PM10, NOx, SO2) through environmental policies to achieve dual environmental benefits. The synergy benefit is used to measure the effectiveness of joint carbon and pollution control efforts. Since carbon emissions, pollutants, and human-generated heat have common sources, this study introduces a carbon-pollution-heat synergy indicator to evaluate urban thermal comfort and equity for residents.

Method

1. Data Collection

Using London as a case study, high-resolution data on carbon emissions, PM2.5, and surface temperature are collected as the prediction targets. Socioeconomic data (e.g., race, age, income) and built environment data (e.g., public transport access, housing vacancy rates, housing types), along with satellite images, are used as predictor variables. The prediction targets and satellite images are image data, while the other variables are tabular, forming a multimodal dataset.

2. Model Development

A conditional generative adversarial network (cGAN) framework is used with a dual-input module to extract features from both image and tabular data, which are then combined. To account for spatial spillover effects, the influence of the 8 surrounding blocks is considered when predicting each block. A graph aggregation method is used to capture the features of neighboring blocks. The model predicts the "carbon-pollution-heat" synergy index for any block.

J. Yao, *Intelligent Exploration in Sustainable Design*,
https://doi.org/10.1007/978-981-96-6743-7_20

3. Model Application

A multi-objective optimization framework is developed based on the synergy prediction model to optimize both governance efficiency and social equity. This method allows for precise control of urban areas, enabling analysis and optimization of key regions (e.g., high synergy with high aging, high density, or low income) to improve governance, environmental benefits, and equity.

Significance
This study develops a multimodal deep learning algorithm to predict urban "carbon-pollution-heat" synergy effects, supporting policy-making, guiding development, and promoting environmental equity.

20.1 Technology Method

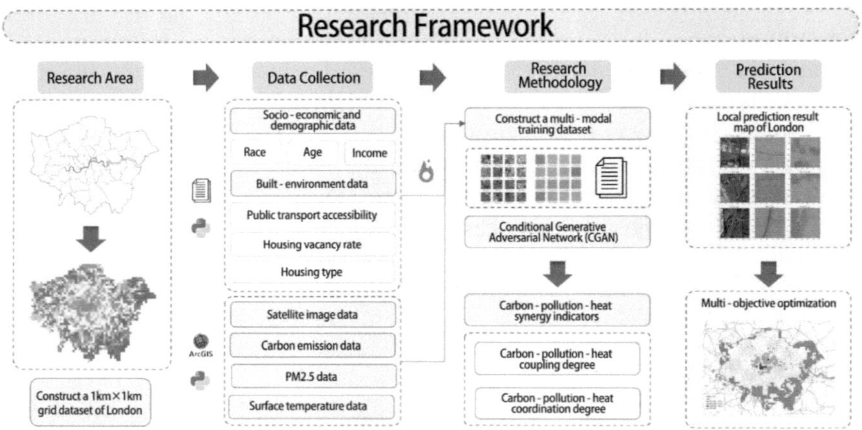

This study creates a 1 × 1 km grid dataset that incorporates socio-economic factors, built environment, carbon emissions, PM2.5, and surface temperature. A Conditional Generative Adversarial Network (CGAN) models the interactions among carbon, pollution, and heat. The analysis of coupling and coordination degrees informs the spatial distribution of these interactions, followed by multi-objective optimization to aid urban planning and environmental management.

20.2 Research Background

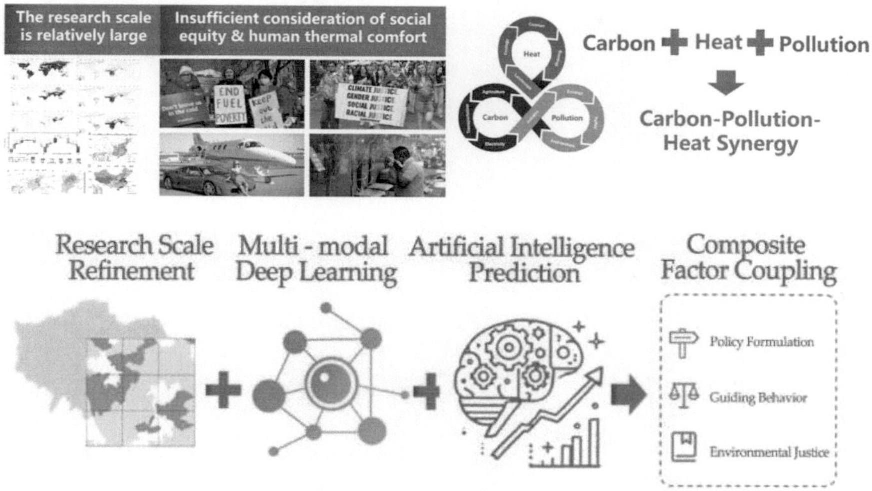

The research background indicates that current studies often overlook social equity and human thermal comfort. This study introduces a carbon-heat-pollution synergy framework using refined research scales, multi-modal deep learning, and AI prediction to analyze the relationships between carbon emissions, thermal environment, and pollution. The findings will aid in policy formulation, behavior guidance, and promoting environmental justice.

20.3 Technology Detail

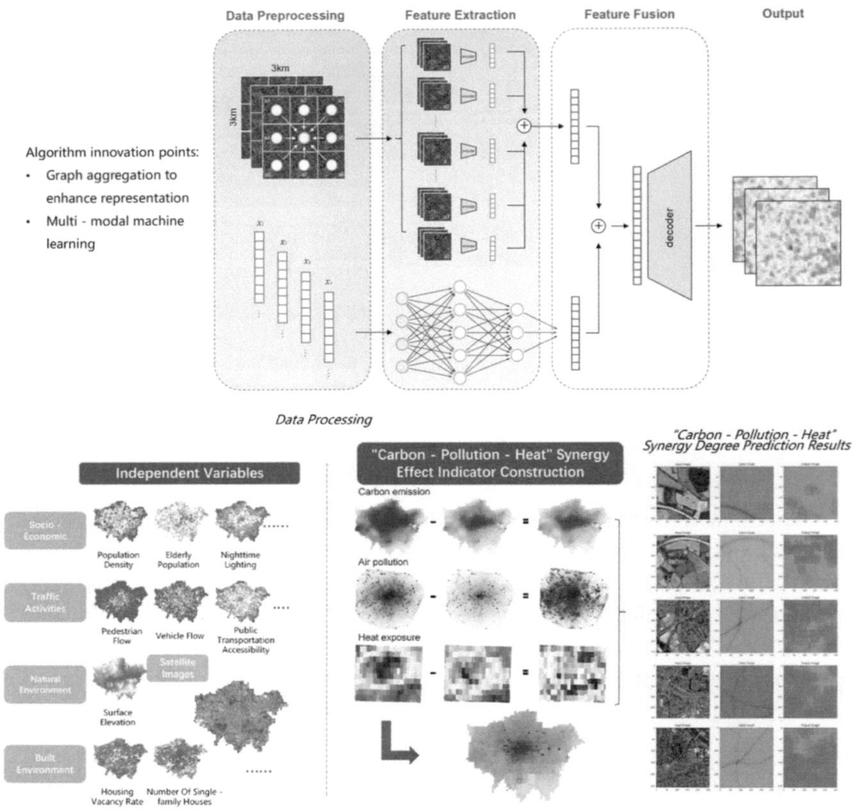

The algorithm combines graph aggregation and multi-modal machine learning to predict the synergy of carbon emissions, pollution, and heat exposure. It processes spatial and socio-economic data, including population density, traffic flow, land use, and satellite imagery. Features are extracted using neural networks and convolutional layers, then fused through a decoder to generate spatial predictions. The results are spatial predictions in the form of carbon-heat-pollution synergy maps, facilitating a comprehensive assessment of urban environmental impacts.

20.4 Research Target

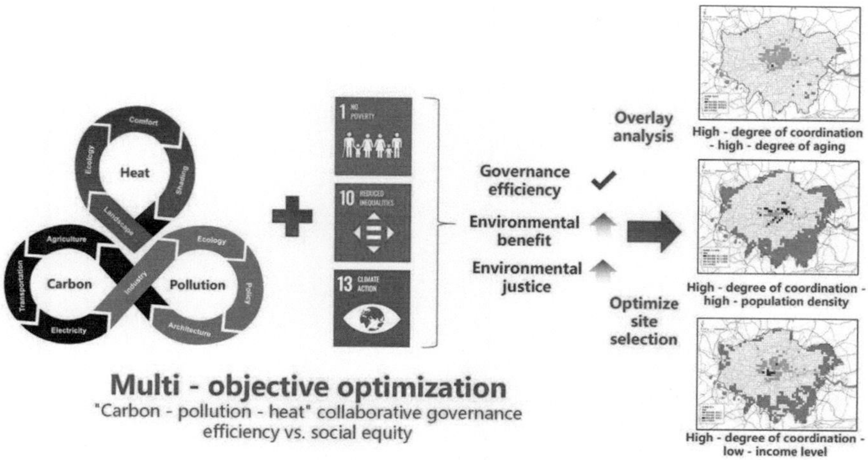

This study examines carbon-pollution-heat governance using a multi-objective optimization approach to balance efficiency, environmental benefits, and social equity. By integrating the United Nations Sustainable Development Goals (SDGs) and employing overlay analysis, it identifies high-coordination areas and optimizes governance site selection based on aging, population density, and income levels to improve environmental justice and policy implementation efficiency.

Chapter 21
Prediction of Net Energy Consumption Potential of District Buildings Based on Generative Adversarial Network Models and Multi-source Image Data

Members
Qin Yuchen, Xie Yuchen, Chen Kaifan, Lei Zhen, Deng Zixuan

Background
By the end of 2020, carbon emissions from the building process accounted for over 51% of China's total emissions. Achieving carbon neutrality is a major challenge for the construction industry and a key step in reaching dual carbon goals. In urban planning, considering both solar PV potential (source) and energy consumption (saving) of buildings can provide a scientific and comprehensive energy demand assessment. However, current methods for predicting PV potential and energy consumption face challenges in efficiency and data availability. This study aims to use multi-source data and generative algorithms to predict the net energy consumption levels of urban blocks.

Method
1. Data Collection

Urban block 3D features and climate parameters affect human activities, which influence building energy consumption and solar PV generation potential, ultimately impacting net energy consumption. Image data for variables like urban texture, building height, function, surface temperature, solar radiation, vegetation index, and wind speed are obtained from Google satellite imagery and remote sensing data. Building energy consumption data is sourced from Southern California's official datasets, and PV potential is calculated using the Ladybug plugin. The difference between energy consumption and PV potential is defined as net energy consumption. These data are preprocessed to create paired image datasets for model training.

2. Model Development

A net energy consumption prediction model is developed using a conditional generative adversarial network (cGAN). Data augmentation is applied, and the model input

J. Yao, *Intelligent Exploration in Sustainable Design*,
https://doi.org/10.1007/978-981-96-6743-7_21

layer is modified to handle multi-channel image data, mapping multiple input images (influencing variables) to a single output (net energy consumption map). The model is fine-tuned to reduce prediction distortions and optimized for the best results.

3. Model Interpretation

Feature importance is analyzed using ablation experiments (feature zeroing analysis). The model's optimal prediction scale is determined by testing urban image slices of different sizes. Finally, Grad-CAM, an explainability algorithm, is used to identify the most influential features for the model's predictions.

Significance
A 400 m range was determined as the optimal study area for predicting the net energy consumption of individual buildings. This study creates a real-world proxy model for urban blocks using multi-source image data, which helps with rapid energy auditing in data-limited scenarios and supports green, low-carbon building design and renovation.

21.1 Technology Method

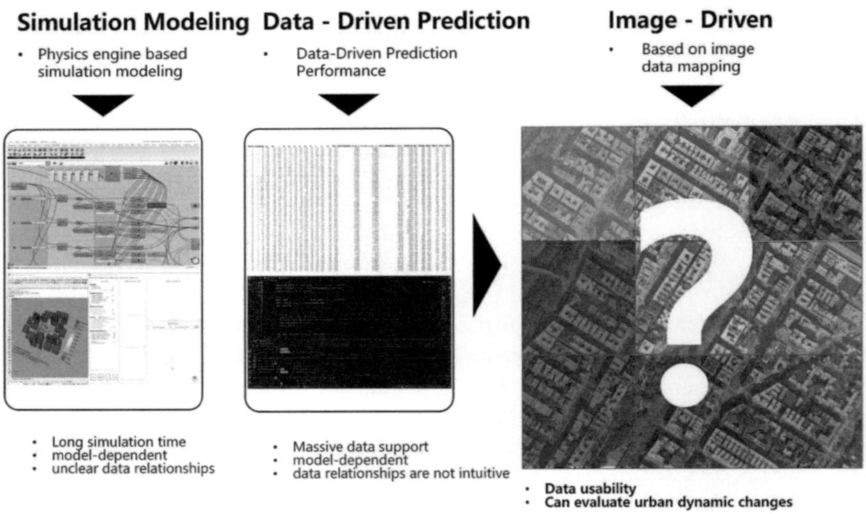

This study analyzes methods for predicting building energy consumption and photovoltaic (PV) output, including physics-based simulations, data-driven approaches, and image-driven techniques. While physics-based simulations are accurate but costly, data-driven methods leverage large datasets but may lack intuitive relationships. Image-driven approaches enhance data usability through image mapping. By integrating multimodal data, this study aims to improve the accuracy and applicability of predictions for building energy consumption and PV output.

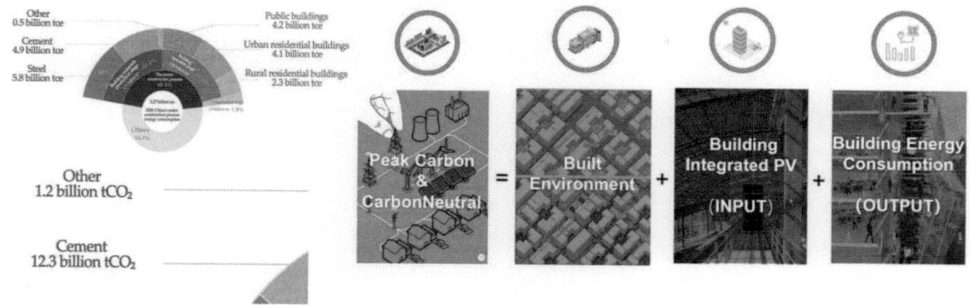

This study emphasizes the building sector's role in carbon emissions and outlines its energy consumption and emissions composition. It suggests achieving carbon peaking and neutrality through optimizing building environments, integrating photovoltaics (BIPV), and predicting energy consumption. By analyzing building morphology, photovoltaic potential, and energy usage, the study offers valuable data for low-carbon building planning.

21.2 Details of Method

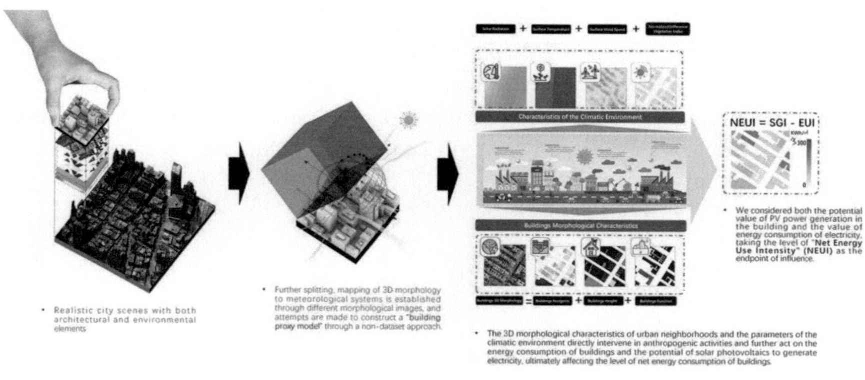

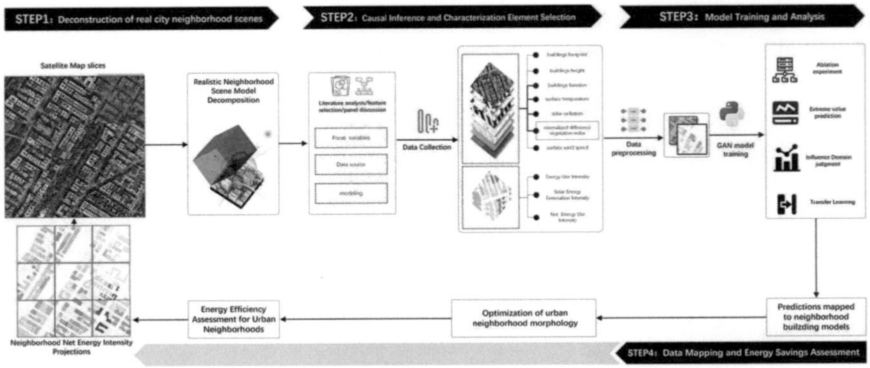

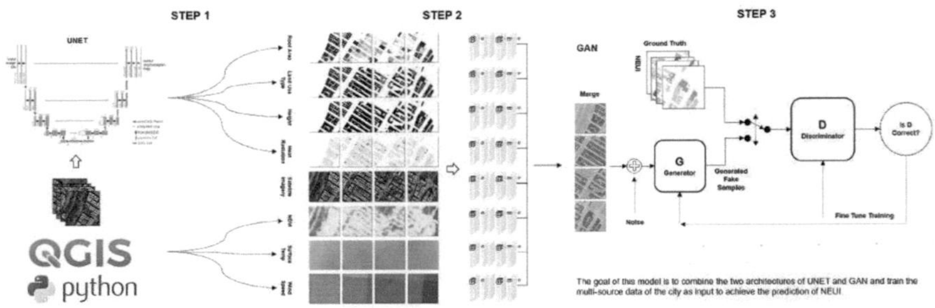

This study uses a multimodal data fusion approach to predict building energy consumption (NEUI) and photovoltaic yield (SGI). It analyzes satellite imagery and GIS data to extract features of urban blocks, then applies causal inference to determine key factors. A deep learning model combining U-Net and GAN is trained for accurate predictions. The results are mapped onto urban block morphology to evaluate their impact on energy conservation and carbon reduction, offering strategies for low-carbon urban planning.

21.3 Research Result

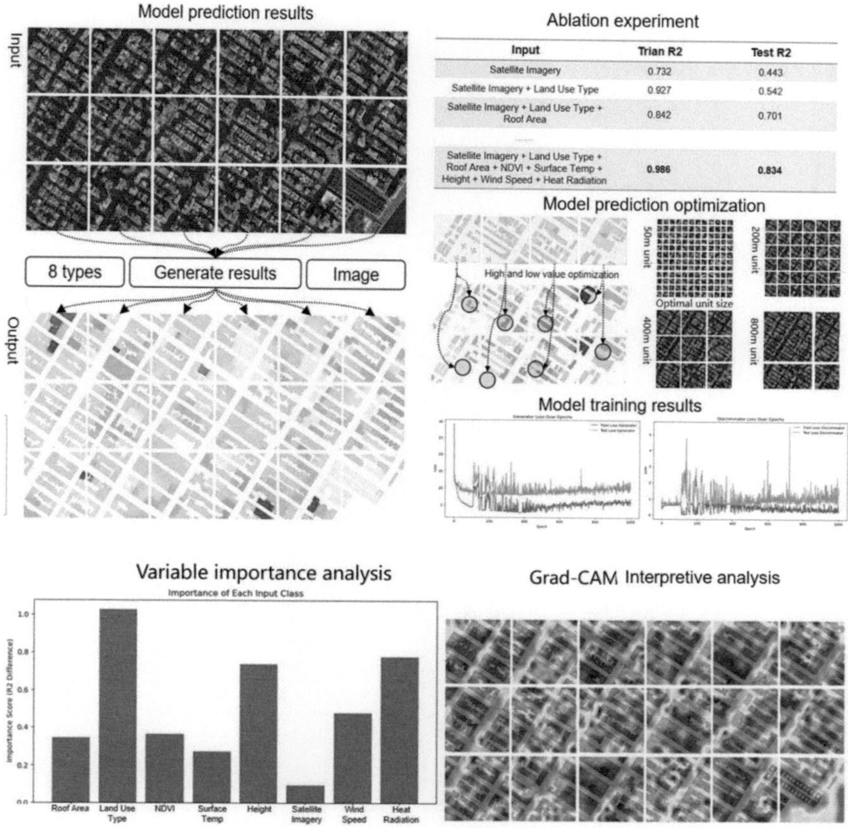

This study predicts building energy consumption and photovoltaic (PV) generation using multimodal data. Ablation experiments showed that factors like roof area, land use type, and meteorological conditions significantly influence prediction accuracy. The optimized model produces precise spatial distribution maps and uses Grad-CAM for interpretability, validating its focus areas. The findings assist in optimizing energy consumption and PV generation in urban blocks, supporting low-carbon urban planning.

Chapter 22
Urban Low-Altitude Aircraft Flight Path Prediction Based on Multi-task Generative Algorithms

Members
Hou Chenfeng, Xu Nuoyan, Ning Qingru, Lin Dianyi, Zhang Yuling

Background
The low-altitude economy is a comprehensive model centered on manned and unmanned civil aircraft, driven by various low-altitude flight activities such as passenger and cargo transport. It promotes the integrated development of related sectors. Future cities will play a key role in this economy. Proper planning of low-altitude airspace is essential for its large-scale development. However, planning faces challenges such as accurately assessing the interaction between aircraft and buildings, the impact of extreme microclimates on flight operations, and noise or interference for residents. This study aims to apply artificial intelligence to assess urban environments and quickly generate low-altitude flight routes.

Method
1. Data Generation

Acquire the geometric model of a small urban block and create a 3D sampling grid. Use Rhino/Grasshopper plugins like Ladybug and Eddy3D to simulate sunlight,

J. Yao, *Intelligent Exploration in Sustainable Design*,
https://doi.org/10.1007/978-981-96-6743-7_22

wind, sightlines, and sound environments in 3D. Batch processing is used to build a multi-physical field database for the city.

2. Model Development

Develop a 3D prediction model using a Diffusion-LSTM architecture. The diffusion model captures multi-scale physical details and handles irregular urban shapes, while the LSTM models sequential relationships between different heights. A 3D CVAE architecture is used for dimensionality reduction, minimizing computational load and memory usage.

3. Trajectory Inference

The pre-trained model can generate real-time 3D information for sunlight, wind, and other physical fields for any urban block. A flight path planning algorithm is created to minimize interference and ensure avoidance, with weighted constraints to optimize route generation based on the city's geometry and physical field data.

Significance

The research shows that the developed 3D multi-physical field prediction model is about 20 times faster than traditional numerical simulations. The flight path planning algorithm outperforms the Dijkstra shortest path algorithm in wind safety and pedestrian interference, effectively avoiding strong turbulence areas. This method offers a valuable reference for drone path planning in complex urban environments.

22.1 Technology Method

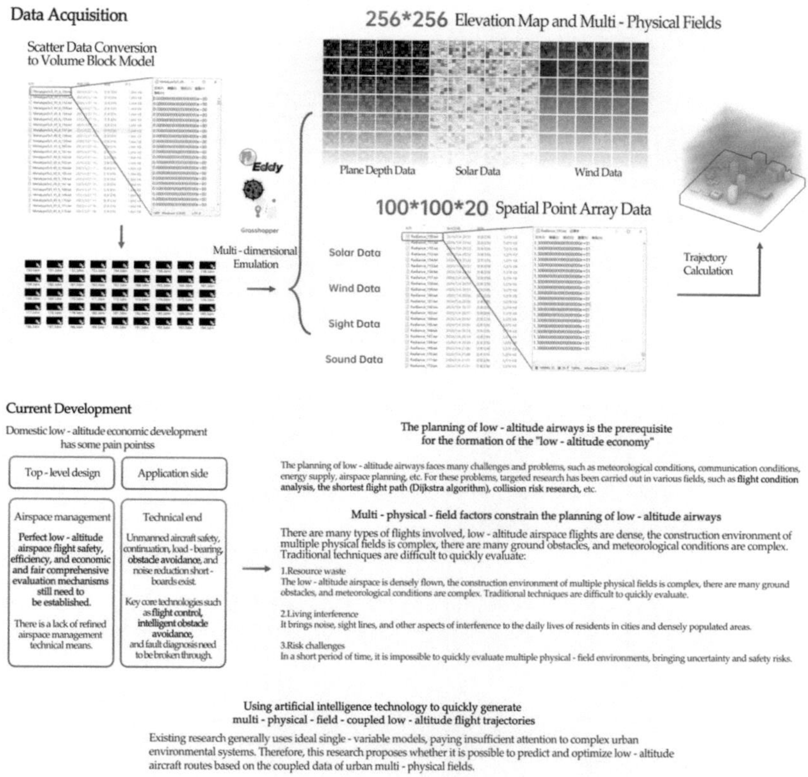

This study addresses low-altitude economy development by proposing an optimization method for flight routes using multi-physical field data. It transforms scatter data into a 3D volume model, incorporating solar radiation, wind fields, visibility, and noise to create a 256×256 elevation grid and a $100 \times 100 \times 20$ spatial point array for flight trajectory calculations. The research identifies challenges in low-altitude airspace planning, such as resource competition and urban interference. By employing artificial intelligence and combining various physical field data, the study seeks to enhance the safety, economic efficiency, and adaptability of low-altitude flight paths, supporting future low-altitude economy growth.

22.2 Research Background

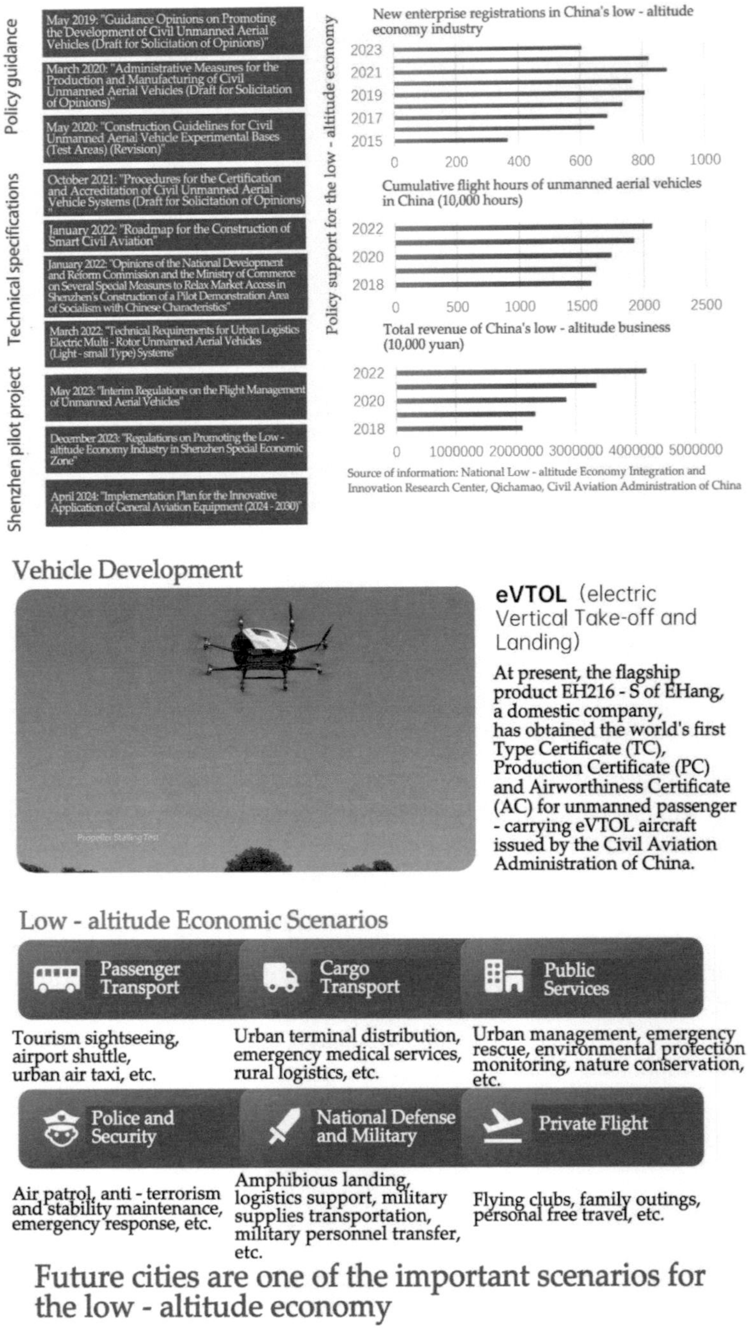

This research background highlights the low-altitude economy, focusing on policy support, technological advancements, and application scenarios. Government initiatives drive growth, particularly for technologies like eVTOL, which enable applications in passenger transport, cargo logistics, public services, and defense. Cities are poised to become essential hubs for developing new urban air mobility models.

22.3 Result of Simulation and Prediction

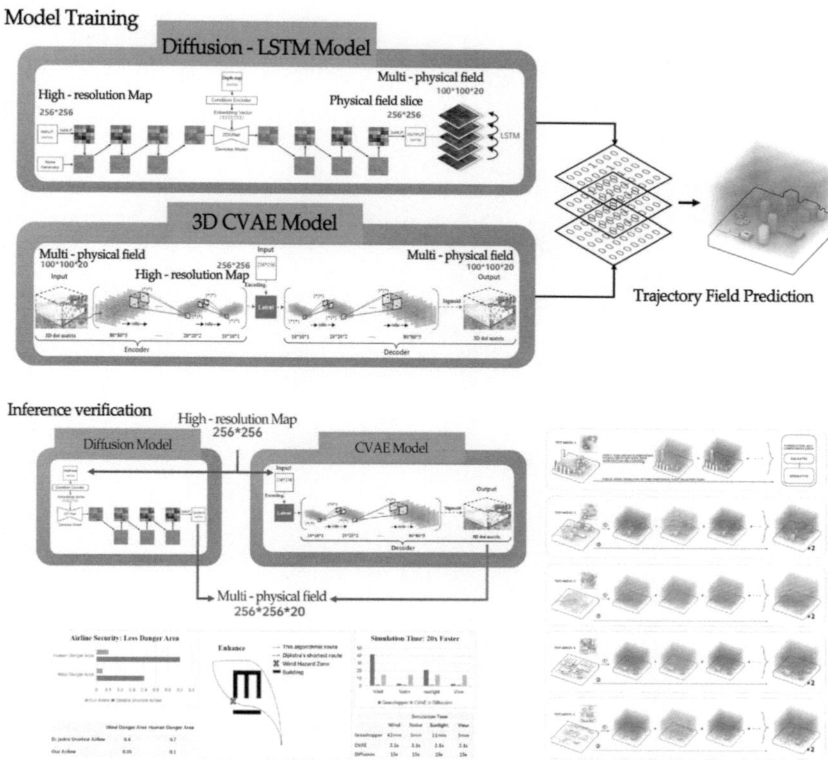

This research employs Diffusion-LSTM and 3D CVAE models for trajectory prediction in complex environments. The Diffusion-LSTM captures spatial-temporal relationships in high-resolution maps, while the 3D CVAE learns latent representations of multi-physical fields. The models enhance safety in low-danger airspace and achieve a 20x faster simulation speed, proving efficient for predicting urban flight paths.

Outlook: The Future of Performance-Driven Design and the New Paradigm of AI in Architectural Education

Reflecting on the past four years of teaching and research, our technological exploration has undergone a significant transformation—from "off-the-shelf application" to independent development. In the initial phase of our teaching practice, we primarily relied on existing AI tools such as machine learning frameworks like TensorFlow and PyTorch, as well as AI plugins in Grasshopper, to help students quickly grasp data-driven architectural optimization methods. The focus at this stage was to familiarize students with the fundamental applications of AI in architectural design, such as accelerating simulations, optimizing form parameters, and analyzing energy consumption and carbon emissions. However, as our research deepened, we gradually realized the limitations of generic AI tools in architectural design. Architectural data is highly complex and multimodal, encompassing various dimensions such as geometric form, environmental impact, structural performance, material properties, and regulatory constraints. Standard AI models, often trained for singular data types, struggle to integrate these multifaceted inputs effectively. Additionally, conventional AI optimization methods largely operate as black boxes, providing designers with numerical results but failing to elucidate the underlying logic or reasoning behind AI-generated solutions. These challenges prompted us to advance toward the customized development of AI technologies that better align with the specific needs of architectural design.

As our research progressed, we gradually developed a series of AI models tailored for architectural applications, shifting our teaching approach from "using AI tools" to "developing AI solutions." We trained graph neural networks to predict architectural performance across different scales, making them adaptable to wind environment analysis, energy consumption prediction, and structural performance optimization. We also explored causal inference methods to uncover deeper relationships between design variables and environmental factors, helping architects understand the rationale behind optimization results. Throughout this process, students transitioned from being passive users of AI-driven optimization to actively considering how AI training methods could be adjusted to better align with architectural reasoning. This shift not

J. Yao, *Intelligent Exploration in Sustainable Design*,
https://doi.org/10.1007/978-981-96-6743-7

only enhanced AI's applicability in architectural design but also provided students with a more comprehensive understanding of both the limitations and potential of AI-driven methodologies, fostering a mindset that integrates data-driven reasoning with design logic.

The future of performance-driven design will not be limited to task-specific AI models but will increasingly evolve toward large-scale pre-trained foundation models. Currently, AI applications in architecture primarily rely on task-oriented modeling, where separate models are trained for specific challenges such as energy prediction, wind environment analysis, and carbon emission calculations. However, this approach has inherent limitations—each new task requires retraining a dedicated model, preventing the formation of a generalizable design optimization framework. In the future, we envision the development of a foundational AI model for architectural performance optimization, one that possesses cross-scale adaptability, multimodal data integration, and self-adaptive reasoning capabilities. Such a model would seamlessly process geometric data, environmental datasets, textual regulations, and other diverse inputs while remaining applicable to design tasks ranging from single buildings to urban-scale planning. Beyond providing optimized design recommendations, this model would also incorporate self-explanatory reasoning to clarify the causal relationships behind its design decisions, enhancing its generalization capabilities and interpretability.

The role of AI in architectural design is also transitioning from a unidirectional computational tool to a genuine design collaborator. Current AI-assisted design processes still follow an "input-output" model: architects input parameters, AI computes the results, and designers adjust accordingly. However, future AI design paradigms will prioritize interactivity, allowing AI to engage in real-time collaboration throughout the design process. We aim for AI to not only provide optimization results but also to explain its decision-making process, enabling architects to understand the rationale behind its suggestions and further refine their design objectives. For instance, when an architect modifies window dimensions, AI should not only calculate changes in lighting, ventilation, and energy consumption but also infer the cascading impact on overall building performance and propose multiple optimization pathways. Moreover, the integration of generative AI and conversational interfaces will enhance AI's role as a design assistant, allowing architects to interact through natural language, gesture-based controls, or immersive AR/VR environments, fostering a more intuitive and collaborative design experience.

Architectural education must further adapt to the paradigm shift brought by AI-driven design and cultivate architects who can think with AI. In our current teaching model, we have already moved beyond basic AI tools to incorporate independent AI development practices, guiding students to integrate data analytics, deep learning, and causal inference into architectural design. Moving forward, we hope to establish a more open AI-driven curriculum, where students not only learn to use AI for architectural optimization but also train their own AI design models and explore their applicability across diverse design tasks. Interdisciplinary collaboration will also play a critical role in future pedagogy, with architecture, computer science, environmental engineering, and urban planning converging to cultivate professionals with a broader, more integrated perspective. Additionally, the development of interactive AI design labs will allow students to experience AI-generated design processes in virtual environments, leveraging data visualization to intuitively grasp AI-generated logic.

In the era of intelligent architectural design, AI will not merely serve as an optimization tool but as a learning and reasoning collaborator that actively analyzes design strengths and interacts deeply with architects. We aspire for AI design methodologies to transcend academic research and find practical applications within the architectural industry, enabling AI to become a truly creative assistant to architects. Future architects will need not only traditional design expertise but also proficiency in data analysis, computational thinking, and an understanding of AI's role in the design process—fully integrating AI as part of their architectural creativity. In this transformation, our workshop is not just an educational experiment but a research platform driving the evolution of AI-assisted design methods. We look forward to seeing more innovations and breakthroughs in this field in the years to come.

This book is not merely a retrospective on our past four years of teaching experiences but also an outlook on the future of AI in architectural design. As AI technology continues to advance, architectural design is undergoing a paradigm shift—from intuition-driven to data-driven, from experience-based to inference-based, and from conventional tools to intelligent collaboration. We hope this book will inspire more researchers, educators, and practitioners to explore the possibilities of AI-assisted design and propel the architectural discipline into a new era of intelligence, efficiency, and sustainability.

MIX
Papier aus verantwortungsvollen Quellen
Paper from responsible sources
FSC® C105338

If you have any concerns about our products,
you can contact us on
ProductSafety@springernature.com

In case Publisher is established outside the EU,
the EU authorized representative is:
Springer Nature Customer Service Center GmbH
Europaplatz 3, 69115 Heidelberg, Germany

Printed by Libri Plureos GmbH
in Hamburg, Germany